THE ALCHEMY OF ALMOST NOTHING

Trace Elements and Their Outsized Influence on Chemical Reactions

Priya Anantharaman

This book is a work of nonfiction. The information contained in this book is presented in good faith and believed to be accurate based on the sources referenced within it. It is intended for general informational and educational purposes only and does not constitute professional, medical, financial, or legal advice. The author and copyright holder disclaim any liability for any loss or damage incurred as a result of the use or application of any information contained in this book.

Contents

A Grain of Salt

In 1924, a batch of table salt went on sale in the state of Michigan that looked, tasted, and poured exactly like every other batch of table salt that had ever been sold there. Nothing about it announced itself. There was no new color, no new smell, no warning label, no fanfare beyond a modest newspaper notice. And yet that shipment of salt would end up doing something that few single products in the history of American public health have matched. It would quietly add fifteen average IQ points to a generation of children in the regions where it was eaten.

The trick had nothing to do with the salt itself. Sodium chloride is sodium chloride, an ancient and unremarkable compound that has flavored food since before recorded history. The trick was a pinch of something else, stirred in at a concentration so low it would have been invisible to any of the eight human senses even if we had a ninth. Iodine, added to salt at roughly one part in fifty thousand, had quietly ended one of the more overlooked medical crises in American history: a belt of iodine-starved soil running through the Midwest, the Pacific Northwest, and the Appalachians, where thyroid glands swelled into visible lumps on the necks of schoolchildren and, more quietly and more permanently, dulled the developing brains of infants whose mothers had never eaten enough of a mineral most of them had never heard of.

This book is about that kind of chemistry. Not the loud kind, with its fireworks and explosions and neon test tubes bubbling over in a high school lab. The quiet kind. The kind measured not in teaspoons but in parts per million, sometimes parts per billion, so small that if you dissolved the entire active ingredient of a chapter's story into an Olympic swimming pool, you might still not be able to detect it without instruments built specifically for the purpose. And yet these vanishingly small quantities of a startlingly short list of elements have, again and again, been the difference between a civilization that starves and one that eats, an engine that knocks itself apart and one that runs for two hundred thousand miles, a computer that exists only as a theoretical curiosity and one that fits in a pocket.

Chemistry has a marketing problem, and it is entirely self-inflicted. Ask most people to picture a chemical reaction and they will picture the dramatic ones: a flame, a color change, a puff of smoke, two liquids combining into a third that fizzes over the beaker's lip. These are the reactions that get filmed for science museum demonstrations, because drama photographs well. But the reactions that have actually reshaped the material conditions of human life tend to look, from the outside, like nothing at all. A gray metal disc sits inside a car's exhaust pipe and converts poison gas into harmless gas, and nothing about its appearance changes while it does this ten thousand times a day for a decade. A wafer of purified silicon, doped with an impurity concentration lower than one atom in a million, becomes capable of switching an electrical current on and off billions of times a second, and it looks, to the naked eye, exactly like a wafer of undoped silicon. A steel knife blade resists rust for a hundred years because someone once melted twelve parts of a strange, blue-gray metal into every hundred parts of iron, and the knife looks, when you pick it up, like any other knife.

The premise of this book is a simple one, though its implications turn out to be enormous: that the story of trace elements is, in large part, the untold story of the modern world. Not the story of gold, or oil, or uranium, whose outsized roles in human affairs get plenty of attention. The story of the also-rans. The elements that show up in the third or fourth decimal place of a material's composition and yet somehow end up controlling everything that happens in the first decimal place. Molybdenum, sitting at the heart of an enzyme that has been quietly converting atmospheric nitrogen into usable form since before there were multicellular organisms to benefit from it. Chromium, discovered almost by accident by a metallurgist trying to stop rifle barrels from corroding, who ended up giving the world stainless steel instead. Cobalt, locked inside a single vitamin molecule so structurally baroque that it took a Nobel Prize-winning crystallographer the better part of a decade just to map its atoms. Lithium, the lightest solid element on the periodic table, discovered in a mineral in the early nineteenth century and largely ignored for a hundred and fifty years until three scientists figured out how to make it carry a electrical charge back and forth between two electrodes billions of times without wearing out, an achievement that eventually put a battery in every pocket on Earth.

Each of these elements, in the quantities that matter, would be difficult to see even under a microscope. And each of them, in the absence of the tiny quantity that matters, would leave behind a world unrecognizably poorer, hungrier, sicker, or darker than the one we actually inhabit.

There is a reason chemists have a special vocabulary for this phenomenon. A catalyst, in the strict technical sense, is a substance that changes the rate of a chemical reaction without being consumed by it. It does not appear in the final tally of reactants and products. It does its work and steps aside, ready to do the same work again, over and over, for as long as it is given the chance. This is a strange kind of power: influence without permanence, control without consumption. A catalyst is not a co-star in the reaction it enables. It is closer to a director, standing just off camera, shaping everything that happens in the frame without ever appearing in it.

Biology discovered this trick first, and it discovered it billions of years before human chemists caught up. Every metabolic pathway inside every living cell on Earth runs on catalysts, mostly proteins called enzymes, and a remarkable number of the enzymes that keep the machinery of life turning have, sitting at their functional core, a single atom of a trace metal. Take that atom away and the enzyme, for all its intricate protein architecture, does nothing. It becomes an elaborate, useless sculpture. Put the atom back and the sculpture wakes up. This is not a metaphor. It is a literal description of what happens inside a nitrogenase enzyme when its cofactor, a cluster of seven iron atoms, nine sulfur atoms, and a single atom of molybdenum, is removed and then restored. The enzyme goes from inert to one of the most consequential catalysts on the planet, responsible, through the chain of organisms it feeds, for a meaningful fraction of the nitrogen atoms in your own body at this very moment.

Human chemistry, once it caught up to biology's trick, ran with it in every direction biology had never bothered to explore, because biology optimizes for what keeps organisms alive and reproducing, not for what keeps automobile engines from knocking or computer chips from melting. And so the twentieth century became, in ways that are rarely narrated as a coherent story, the century of the trace element. It was the century when a German chemist figured out how to use an iron catalyst laced with potassium and aluminum oxide to pull nitrogen directly from the air and turn it into fertilizer, ending the centuries-old dependence on manure and guano and averting a famine that demographers had genuinely believed was coming. It was the century when three physicists at a New Jersey research laboratory discovered that a sliver of germanium, treated with the right trace impurities in exactly the right places, could amplify an electrical signal without the bulk, fragility, and heat of a vacuum tube, and in doing so lit the fuse on the entire digital age. It was the century when a British metallurgist, trying to solve a mundane problem about rifle barrels wearing out too fast, stumbled onto an alloy that would not rust, and unknowingly launched an industry that today shapes everything from surgical instruments to skyscraper facades. It was, less happily, also the century when an American engineer discovered that a few grams of a lead compound added to a gallon of gasoline could quiet an engine's knock, and in doing so set in motion a slow-moving public health catastrophe whose consequences took seventy years and one obstinate geochemist to fully understand and finally undo.

These stories, and others like them, form the chapters of this book. Each one begins with an element so minor, in terms of raw quantity, that it barely deserves the label "ingredient." A pinch, a trace, a whisper, a rounding error. And each one ends with a demonstration, drawn from real historical and scientific record, of just how much leverage a whisper can exert on the world, if it is whispered into precisely the right place.

I have tried, in telling these stories, to resist two temptations that tend to ruin popular science writing. The first is the temptation to flatten every discovery into a single triumphant eureka moment, as though Harry Brearley looked at a crucible of molten chromium steel and instantly understood he had changed the future of metallurgy forever. He did not. He noticed that a discarded sample had not rusted the way the rest of his failed rifle-barrel experiments had, and it took him months of quiet, unglamorous persistence, plus a considerable amount of professional indifference from the people around him, before anyone thought to make a knife out of it. The real history of trace element chemistry is full of dead ends, professional rivalries, years of tedious purification work, and discoveries that sat unrecognized for decades before someone understood their significance. That messiness is not a flaw in the story. It is the story.

The second temptation is to pretend that trace elements are uniformly heroic, quiet saviors working behind the scenes to make our lives better. They are not. The same quality that makes a trace element powerful when it is present, its ability to exert outsized effects at vanishingly small concentrations, also makes it dangerous when it is present in the wrong place, in the wrong form, or in slightly too high a dose. Lead is the starkest example in this book, but it is not the only one. The line between a trace element that quietly builds and a trace element that quietly poisons is often a matter of which enzyme it binds to, which biological pathway it interferes with, and how stubbornly it resists being flushed back out of the body once it has entered. Chemistry does not hand out moral character along with atomic number. It hands out reactivity, and what a society does with that reactivity, whether it regulates it, studies it, profits from it, or ignores the warning signs for seventy years because a powerful industry finds the warnings inconvenient, is a human story layered on top of a chemical one.

So this is a book about very small things that do very large work, told through the people who found them, harnessed them, misused them, and, in some cases, spent the second half of their careers trying to undo the damage caused by the first half. It moves from the nitrogen-starved soils of the early twentieth century to the exhaust pipes of modern automobiles, from the thyroid glands of Midwestern schoolchildren in the 1920s to the cathodes of the lithium battery in whatever device you may be reading this on. It is not a comprehensive periodic table of every trace element that has ever mattered. Such a book would run to several volumes and would not be much fun to read. Instead, it is a selection, chosen for the strength of its stories and the clarity with which each one illustrates the book's central claim: that the history of the modern material world is, to a surprising degree, a history written in the margins, in the parts per million, in the elements too small to see and too important to ignore.

Consider it a grain of salt. Or rather, consider it the pinch of something else stirred into that grain, the part you cannot taste and would never notice, doing more work than everything else on the shelf combined.

Chapter 1

The Tyranny of the Tiny

How much is a part per million, really?

Picture a large lecture hall packed with a million people, shoulder to shoulder, filling every seat and spilling into the aisles. Now imagine that exactly one of those million people is wearing a red shirt while everyone else wears identical gray. That single red shirt, lost somewhere in a sea of a million gray ones, is a part per million. It is the concentration at which molybdenum typically appears in agricultural soil, the concentration at which many of the additives that transform steel, semiconductors, and catalytic converters do their work, and the concentration at which the human body regulates several of the trace minerals it cannot survive without.

Now shrink the stadium. Imagine not a million people but a billion, a crowd so large it would require roughly three New York Cities standing shoulder to shoulder, and somewhere in that vast ocean of gray shirts, a single red one. That is a part per billion, the scale at which some of the rarest and most consequential biological catalysts operate, and the scale at which certain environmental contaminants, undetectable to any human sense, can still disrupt the hormonal systems of entire ecosystems.

These are not numbers that intuition handles well. Human beings evolved to reason about quantities we could count on our fingers, herds we could see grazing across a valley, harvests we could measure in baskets. We did not evolve to reason about one part in a million, because for the overwhelming majority of our species' history, nothing observable to the naked eye depended on quantities that small. And yet the story this book tells, again and again, is the story of quantities exactly that small turning out to be the hinge on which enormous outcomes swing. A crop that starves a village or feeds it. An engine that seizes after ten thousand miles or purrs along for two hundred thousand. A computer chip that exists only as a laboratory curiosity or one that fits, along with several billion of its cousins, into the device most people now carry in a pocket.

Chemists have a word for the strange kind of influence exercised by substances present in these vanishing quantities. They are catalysts, or promoters, or dopants, or trace elements, depending on the exact discipline and context, but the underlying phenomenon is the same across all of these labels. A very small amount of one substance, added to a much larger system, changes the behavior of that system out of all proportion to its own quantity. The substance itself is usually not consumed and does not become part of the final product in any obvious way. It sits at the edges, or buried deep within the structure, doing its work quietly and then, in many cases, doing it again, and again, essentially forever.

To understand why this works, and why it does not always work, it helps to start with the basic architecture of a chemical reaction, and then to understand the several distinct ways a trace element can insert itself into that architecture and take control.

THE SPEED OF THINGS THAT WANT TO HAPPEN

Every chemical reaction you have ever heard of, from the rusting of an iron nail to the combustion of gasoline in a car engine to the metabolic reactions keeping you alive as you read this sentence, has to clear an energy hurdle before it can proceed. Chemists call this hurdle the activation energy, and it exists because breaking the chemical bonds holding reactant molecules together requires an initial input of energy, even if the overall reaction, once it gets going, releases far more energy than it consumed.

Think of a boulder resting in a shallow depression at the top of a hill. The boulder wants, in the sense that physics gives meaning to wanting, to roll down the hill, because the bottom of the hill represents a lower energy state than the top. But the boulder cannot simply roll away on its own, because the lip of the depression it sits in is slightly higher than the boulder's current position. Someone or something has to give it an initial push, just enough to get it up and over that lip, after which gravity takes care of the rest.

The activation energy of a chemical reaction plays exactly the role of that lip. Nitrogen gas in the atmosphere, for example, is desperately eager, thermodynamically speaking, to combine with hydrogen and form ammonia, a reaction that releases a substantial amount of energy once it occurs. And yet the nitrogen in the air around you right now is not spontaneously turning into ammonia, because the two nitrogen atoms in an N2 molecule are held together by one of the strongest chemical bonds in nature, a triple bond that requires an enormous initial energy input to break. The boulder wants to roll, but the lip is very high indeed.

A catalyst works by giving the reacting molecules an alternative path over a lower lip. It does not change where the boulder starts or where it ends up. It does not change the fact that nitrogen and hydrogen, once combined into ammonia, sit at a lower energy state than they did as separate gases. What it changes is the shape of the hill in between, carving a gentler slope that requires less of an initial push to get the boulder moving. In the specific and enormously consequential case of industrial ammonia production, that gentler slope is provided by the surface of an iron catalyst, which gives nitrogen molecules a place to land, weakens the grip of their triple bond just enough for it to be pried apart, and then offers hydrogen atoms a nearby perch from which to bond with the newly liberated nitrogen atoms. The iron itself is not consumed in this process. It ends the reaction exactly as it began, ready to catalyze the next batch of nitrogen and hydrogen molecules that happen to drift across its surface.

This is the first and most fundamental role a trace element plays in the chemistry of the modern world: lowering an energy barrier that would otherwise make a useful reaction impossibly slow, or in some cases, so slow as to be indistinguishable from not happening at all.

THE SURFACE IS WHERE EVERYTHING HAPPENS

A crucial detail, easy to overlook, is that most industrial and biological catalysis happens not throughout the volume of a catalytic material but specifically at its surface, the boundary layer where the catalyst meets the reactants trying to pass over it. This is why so much of catalyst engineering, across a startling range of applications, boils down to a single obsessive question: how do you maximize the amount of surface area exposed to the reaction, using the smallest possible quantity of what is often an expensive or rare material?

The platinum, palladium, and rhodium inside a modern catalytic converter illustrate this principle vividly. These metals are precious in the literal sense, traded on commodity markets alongside gold and silver, and a catalytic converter loaded with a solid brick of any of them would be prohibitively expensive for a mass-market vehicle. Instead, engineers coat a honeycomb-shaped ceramic support structure with an extremely thin layer of these metals, sometimes measured in single-digit micrometers, spread across a surface area that, if unfolded and laid flat, would cover an area comparable to a football field. The actual mass of precious metal inside a typical catalytic converter is often just a few grams, an amount that would fit easily in the palm of your hand, and yet that small mass, spread thin across a vast folded surface, is enough to process the exhaust of an engine burning through gallons of fuel every hour, converting toxic carbon monoxide, unburned hydrocarbons, and nitrogen oxides into carbon dioxide, water vapor, and harmless nitrogen gas.

The same logic, refined over three and a half billion years of evolution rather than a century of industrial engineering, governs the metal centers of biological enzymes. The molybdenum atom inside the nitrogenase enzyme, discussed at length in the next chapter, is not distributed throughout the enzyme's bulk. It sits at a single, precisely constructed active site, surrounded by a scaffold of iron and sulfur atoms that chemists call the iron molybdenum cofactor, or FeMoco for short, a structure so complex that its exact atomic arrangement was not fully resolved by crystallographers until decades after the enzyme itself had been studied. The rest of the enzyme, the vast majority of its considerable bulk, exists essentially to hold that one small metal cluster in exactly the right position, at exactly the right angle, exposed to exactly the right molecules, so that the trace metal at its heart can do the one specific job the entire structure has been built to enable.

THREE WAYS A TRACE ELEMENT TAKES CONTROL

Across the stories told in this book, trace elements exert their outsized influence through three broadly distinguishable mechanisms, though in practice these often overlap and reinforce one another.

The first is direct catalysis, the mechanism already described, in which the trace element itself provides the surface or active site where a chemical transformation occurs. Molybdenum in nitrogenase, iron in the Haber process, platinum and rhodium in a catalytic converter, all belong to this category. Remove the trace element and the reaction either does not happen at all or happens so slowly as to be practically useless.

The second is promotion, a subtler and in some ways more surprising mechanism, in which a trace element does not catalyze the reaction directly but instead improves the performance of a separate catalyst by altering its physical or electronic structure. The potassium added to the iron catalyst in the Haber process is the textbook example. Potassium atoms do not themselves grab hold of nitrogen molecules or hydrogen molecules. Instead, sitting on the iron surface in tiny quantities, they donate electron density to nearby iron atoms, which in turn makes those iron atoms more effective at weakening the nitrogen triple bond. Aluminum oxide, added alongside the potassium, plays a different promotional role entirely, acting as what chemists call a structural promoter, physically preventing the iron catalyst particles from clumping together into larger, less efficient masses as the reaction proceeds at high temperature for months or years on end inside an industrial reactor. Neither the potassium nor the aluminum oxide does the catalytic work itself. Both make the actual catalyst, the iron, dramatically better at its job.

The third mechanism is structural alteration, in which a trace element is incorporated not at a reactive surface but throughout the bulk structure of a material, changing its fundamental physical or electrical properties. This is the mechanism behind the chromium in stainless steel, which forms a thin, invisible, self-healing layer of chromium oxide across the surface of the alloy, a passivation layer that physically blocks oxygen from reaching the iron underneath and rusting it. It is also the mechanism behind the trace impurities, phosphorus and boron among them, deliberately introduced into otherwise near-perfectly pure silicon crystals to create the semiconductor devices that underpin every piece of modern electronics, a process called doping that will occupy an entire chapter later in this book. In both cases, the trace element is not catalyzing a reaction in the traditional sense. It is rewriting the rules of the material it has been added to, at a structural or electronic level, using a quantity so small that the material's bulk composition remains, on paper, almost unchanged.

WHY SO LITTLE CAN DO SO MUCH

There is a temptation, when first encountering these facts, to treat them as curiosities, chemistry trivia fit for a dinner party but not much more. This temptation should be resisted, because the pattern that recurs throughout this book, small quantities exerting large and disproportionate effects, is not an isolated set of coincidences. It reflects something close to a general principle about how complex systems behave, one that shows up again and again across biology, materials science, and industrial chemistry, and understanding it changes how you see a remarkable range of phenomena in the world around you.

The principle is this: in any sufficiently complex system, whether it is a metabolic pathway, a crystal lattice, or an industrial reactor, there are typically a small number of points where the system's overall behavior is unusually sensitive to change, surrounded by a much larger number of points where change makes comparatively little difference. Engineers sometimes call these sensitive points bottlenecks or rate-limiting steps. Biologists call them regulatory nodes. Whatever the label, the underlying mathematics tends to be similar: influence in complex systems is rarely distributed evenly. It concentrates at chokepoints, and a small intervention applied precisely at a chokepoint can produce effects wildly out of proportion to the size of the intervention itself, while the same amount of material or energy applied somewhere else in the system might produce almost no effect at all.

Trace elements, it turns out, are extraordinarily good at finding and occupying these chokepoints, for reasons rooted in basic atomic physics. Transition metals in particular, the broad family of elements that includes iron, cobalt, molybdenum, chromium, and platinum, possess an unusual electronic structure, with partially filled d-orbitals that allow them to form and break chemical bonds with a flexibility most other elements cannot match. This flexibility makes them exceptionally well suited to sit at the chokepoints of chemical reactions, forming temporary bonds with reactant molecules, weakening the bonds those molecules would otherwise need to break on their own, and then releasing the transformed products so the metal center can repeat the cycle with the next molecule that arrives. A single molybdenum atom inside a nitrogenase enzyme can participate in the fixation of many nitrogen molecules over the course of its functional lifetime, each cycle taking only a fraction of a second, which means that the enzyme's overall productivity, and by extension a meaningful portion of the nitrogen cycle that sustains life on this planet, rests on the electronic peculiarities of an element present in soil at concentrations too low for most other purposes to even notice.

A REVERSE READING OF THE PERIODIC TABLE

Chemistry students typically encounter the periodic table as a chart organized by increasing atomic number, memorizing groups and periods, valence electrons and electronegativity trends. This is a perfectly reasonable way to learn the fundamentals, but it obscures a different and in some ways more useful way of reading the same information, one this book leans on throughout: reading the periodic table by leverage rather than by abundance.

Read this way, the periodic table sorts itself into two rough camps. On one side sit the elements of bulk, the ones measured by the ton rather than the milligram: iron, aluminum, carbon, oxygen, silicon in its untouched form, the raw material of structures, vehicles, and infrastructure. These elements provide mass, strength, and volume, and civilizations have always needed enormous quantities of them to build anything at scale.

On the other side sit the elements of leverage, a much smaller and stranger group, most of them transition metals or metalloids, that matter not because of how much of them exists but because of how precisely and cleverly they can be deployed in tiny amounts to control the behavior of the bulk elements around them. A skyscraper's steel frame is made overwhelmingly of iron, but the small fraction of chromium mixed into its stainless components is what keeps the exposed surfaces from corroding into rust streaks within a decade. A car's engine block is made overwhelmingly of steel and aluminum, but the few grams of platinum, palladium, and rhodium in its exhaust system are what keep that engine from being, by regulatory standards, illegal to drive. A modern smartphone contains, by weight, mostly plastic, glass, and aluminum, but the trace quantities of lithium in its battery and the trace impurities doped into its silicon processor are what make it a smartphone rather than an inert brick.

MEASURING WHAT CANNOT BE SEEN

A fair question, and one worth pausing on before moving further into this book's individual stories, is how scientists actually measure a quantity as small as a part per million or a part per billion in the first place, since these concentrations sit well below anything the naked eye, or even a good optical microscope, could ever hope to detect. The answer lies in a family of analytical techniques developed across the twentieth century specifically to solve this problem, and understanding the broad shape of how they work helps explain why so many of the discoveries in this book took decades to make, rather than years.

Atomic absorption spectroscopy, one of the workhorse techniques of trace element chemistry, exploits a simple physical fact: atoms of a given element absorb light at very specific, characteristic wavelengths, determined by the precise arrangement of electrons around that element's nucleus. By vaporizing a sample and passing a beam of light tuned to a particular element's characteristic wavelength through the resulting vapor, researchers can measure how much of that light gets absorbed, and from that absorption, calculate with considerable precision how many atoms of the target element were present in the original sample, even when those atoms number only a handful for every million or billion atoms of everything else. Mass spectrometry, a complementary and in many respects even more sensitive technique, takes a different approach entirely, ionizing a sample and then sorting the resulting charged particles by their mass to charge ratio, allowing researchers to identify and quantify individual elements, and in some cases individual isotopes of a single element, with a precision that can reach into the parts per trillion range under favorable laboratory conditions.

These techniques did not exist, at least not in a form sensitive enough to be broadly useful, for most of the twentieth century's early decades, which helps explain why so many of the discoveries recounted in this book, cobalt's role in vitamin B12 among them, took as long as they did. Researchers working before the widespread availability of sensitive trace analysis instruments were, in a very real sense, working partially blind, forced to infer the presence and behavior of trace elements indirectly, through their biological or chemical effects, rather than measuring them directly. The steady improvement of analytical instrumentation across the twentieth century, itself a largely uncelebrated scientific achievement in its own right, quietly enabled much of the more visible progress chronicled in the chapters that follow.

This book is a tour of the elements of leverage. It tells the stories of the people who discovered how to deploy them, the discoveries and accidents that revealed their power, and the consequences, both triumphant and catastrophic, that followed once humanity learned to whisper a few parts per million into precisely the right place. It begins, in the next chapter, with an element that biology discovered billions of years before any human chemist thought to look for it: molybdenum, sitting at the heart of the enzyme that first taught life on Earth how to eat the air itself.

Chapter 2

The Nitrogen Trap

Nearly four out of every five molecules in the air you just inhaled were nitrogen, and every one of them slipped back out on your next exhale without your body doing anything with them at all.

This is one of the odder facts about the atmosphere we live inside. Nitrogen is the most abundant element in the air by a wide margin, and it is also one of the essential building blocks of life, a required ingredient in every protein, every strand of DNA, every molecule of chlorophyll that captures sunlight for a growing plant. And yet, despite swimming in an ocean of it, animals and plants cannot simply pull nitrogen gas out of the air and use it. The nitrogen molecule, two nitrogen atoms bound together by one of the strongest chemical bonds known to chemistry, is essentially inert as far as biology is concerned, sailing through lungs and leaves without reacting with anything.

This creates a strange kind of scarcity in the midst of abundance, and for most of the history of agriculture, it was the single greatest constraint on how much food a given patch of land could produce. A field could have plenty of water, plenty of sunlight, and reasonably fertile soil, and still yield disappointing harvests year after year, because the plants growing in it could not access the one element they needed most, despite that element making up seventy eight percent of the sky above them. Farmers across cultures and centuries developed workarounds without fully understanding the underlying chemistry: rotating in nitrogen-fixing legume crops, spreading manure, and, by the nineteenth century, importing vast quantities of bird droppings from islands off the coast of Peru, a substance called guano that briefly became one of the most strategically valuable commodities on Earth precisely because it was rich in a form of nitrogen plants could actually use.

Somewhere, though, on a much older timescale, nature had already solved this problem. It had done so not through any process a human engineer would have designed, but through the patient trial and error of evolution, and the solution it arrived at hinges on an element that, on its own, seems like an unlikely hero: molybdenum, a hard, silvery transition metal that occurs in the earth's crust and in ordinary soil at concentrations so low they barely register on a standard chemical analysis.

A DUTCH MICROBIOLOGIST AND A JAR OF DIRT

The first hint that something unusual was going on came at the turn of the twentieth century, from the work of a Dutch microbiologist named Martinus Beijerinck. Beijerinck was studying a free-living soil bacterium called Azotobacter chroococcum, and he noticed something that should not have been possible according to the biology of the time: this bacterium could grow rapidly in a medium that contained no fixed nitrogen source whatsoever, no ammonia, no nitrate, none of the forms of nitrogen that every other known organism required. The bacterium appeared to be pulling nitrogen directly out of the air, converting the inert gas into a form its own cells could use to build proteins and genetic material.

This was, in effect, the discovery of biological nitrogen fixation, though it would take decades of subsequent research to understand the mechanism behind it. Researchers working in the years after Beijerinck's initial observation began systematically testing which trace minerals, added to the growth medium in small quantities, might be enabling this remarkable feat. They tried molybdenum, in the chemical form of sodium molybdate, and they tried vanadium, in the form of sodium metavanadate, and what they found reshaped the trajectory of agricultural science for the rest of the century. Molybdenum stimulated the nitrogen-dependent growth of Azotobacter by a factor of roughly three compared to control samples that lacked it. Vanadium helped as well, though less dramatically. The conclusion researchers eventually reached, confirmed across studies of other nitrogen-fixing microorganisms including species of Clostridium and Rhodospirillum, was unambiguous: molybdenum was essential for biological nitrogen fixation, a trace metal without which an entire category of microbial metabolism simply could not function.

BEIJERINCK, THE MAN WHO NAMED VIRUSES TOO

Beijerinck's contribution to trace element biology, significant as it was, represents only one part of a genuinely remarkable scientific career, and a brief detour into the rest of it helps illustrate the kind of broad, foundational curiosity that characterized so many of the early microbiologists whose work made later, more specialized discoveries possible. Beijerinck is also widely credited as one of the founders of virology, having conducted pioneering experiments on tobacco mosaic disease in the 1890s that led him to conclude the infectious agent responsible was smaller than any known bacterium and capable of replicating only within a living host cell, a genuinely novel category of biological entity that Beijerinck termed a contagium vivum fluidum, or contagious living fluid, an awkward phrase at the time that nonetheless anticipated much of what would later be formalized as the modern concept of a virus.

Beijerinck's career, spanning both the discovery of an entirely new category of infectious agent and the identification of molybdenum's essential role in nitrogen fixation, illustrates a pattern common among the foundational figures of early twentieth century microbiology: working in an era before scientific research had become as narrowly specialized as it is today, these researchers moved comparatively freely across what would now be considered distinct subfields, bringing the same rigorous, patient, observational approach to whatever unexplained biological phenomenon happened to capture their attention. Beijerinck's insistence on careful, controlled experimentation, varying one factor at a time against an appropriate control group, a methodological standard that seems unremarkable today but was still being actively established as a norm within biology during his working lifetime, proved just as valuable in unraveling the mystery of nitrogen fixation as it did in identifying the entirely novel category of pathogen responsible for tobacco mosaic disease.

INSIDE THE MOST IMPORTANT CLUSTER OF ATOMS YOU HAVE NEVER HEARD OF

Understanding why molybdenum plays this role required identifying and studying the enzyme responsible for nitrogen fixation itself, a protein complex that biochemists eventually named nitrogenase. Nitrogenase turns out to be a strikingly complicated piece of biological machinery, and at its functional core sits a metal cluster of genuine structural beauty, one that took researchers years of painstaking crystallographic work to fully resolve.

The active site of the most common form of nitrogenase is called the iron molybdenum cofactor, universally abbreviated by researchers as FeMoco, and its composition reads like a small chemistry puzzle: seven iron atoms, nine sulfur atoms, a single molybdenum atom, one additional light atom whose exact identity took researchers years to pin down, and a small organic molecule called homocitrate holding the whole structure together. This cluster is, by some measures, the largest metal cofactor known in biology, an intricate cage-like arrangement of atoms that provides the specific chemical environment needed to grab hold of an inert nitrogen molecule and, through a sequence of electron transfers requiring a substantial input of cellular energy, break its formidable triple bond and combine it with hydrogen to produce ammonia.

The overall reaction, written out in the chemical shorthand biochemists use, requires nitrogen gas, hydrogen ions, electrons, and a striking sixteen molecules of ATP, the cell's primary energy currency, to produce two molecules of ammonia along with a molecule of hydrogen gas as an unavoidable byproduct. Sixteen molecules of ATP is an enormous energetic investment by cellular standards, and it reflects just how stubbornly the nitrogen triple bond resists being broken, even with the considerable catalytic assistance the FeMoco cluster provides. Without that cluster, and specifically without the single molybdenum atom sitting near its center, the reaction simply does not proceed, no matter how much cellular energy an organism is willing to throw at the problem.

There is a genuine irony buried in the details of this cofactor's structure. Among the many molybdenum-dependent enzymes identified across the tree of life, more than fifty of them by current counts, nitrogenase is actually something of an outlier, lacking a chemical feature called molybdopterin that most other molybdenum enzymes rely on to hold their metal center in place. Nature appears to have arrived at the use of molybdenum in nitrogenase through an entirely separate evolutionary pathway from the one it used for molybdenum's other enzymatic roles, a reminder that evolution does not design elegant, unified solutions so much as it stumbles repeatedly toward whatever happens to work, using whatever raw materials happen to be available.

THE DECADES IT TOOK TO SEE THE CLUSTER CLEARLY

Determining the precise atomic structure of FeMoco proved to be one of the more protracted structural biology campaigns of the twentieth century, a fitting parallel to the extended crystallographic effort Dorothy Hodgkin would later invest in mapping vitamin B12's cobalt-centered structure, described in a later chapter of this book. Researchers first obtained a low resolution structural picture of the nitrogenase enzyme in the 1990s, sufficient to establish the broad outline of the cluster's iron, sulfur, and molybdenum content but not yet sufficient to resolve every atom with full confidence. It was not until 2002 that researchers, employing considerably improved X-ray crystallography techniques, finally identified the previously mysterious light atom sitting at the very center of the cluster as a carbon atom, an unusual and, at the time, genuinely unexpected structural feature that had eluded confident identification for years despite the cluster's central importance to global nitrogen cycling.

This extended timeline, spanning multiple decades and generations of steadily improving crystallographic technology, echoes a pattern this book will note repeatedly in its later chapters as well: even once a trace element's general importance to a given biological or chemical process has been firmly established, mapping the precise atomic details of how that element actually accomplishes its function often requires additional decades of dedicated, technically demanding research, conducted by scientists willing to spend the greater part of a career refining a single structural picture down to the level of individual, correctly placed atoms.

THE ELEMENT THAT SOILS RUN SHORT OF

For most of human agricultural history, nobody needed to think about molybdenum, because nobody understood that it was the invisible linchpin behind a process, biological nitrogen fixation, that quietly enriched the soil in fields planted with legumes like clover, alfalfa, and beans. Farmers observed the beneficial effect of rotating these crops into their fields without understanding the mechanism, attributing the improved yields that followed to a general notion of the soil "resting" rather than to a specific microbial partnership between legume roots and molybdenum-dependent nitrogen-fixing bacteria living in nodules on those roots.

It was only in the twentieth century, once the biochemistry of nitrogenase became better understood, that researchers realized molybdenum deficiency in soil was a genuine and significant agricultural problem in its own right, distinct from the more commonly discussed deficiencies of nitrogen, phosphorus, and potassium that fertilizer companies had built entire industries around. Because molybdenum is required in such minute quantities, plant tissue typically contains it at concentrations of roughly half a part per million on a dry matter basis, a level far lower than the concentrations required for other essential nutrients, it had simply not registered as a priority for agricultural scientists focused on the elements needed in bulk.

This changed as researchers began studying nitrogen-poor tropical soils in detail, and the results were striking. Studies conducted in Central America, including research near sites considered sacred in

Maya mythology, found that molybdenum, not phosphorus as had long been assumed, was frequently the limiting factor controlling the activity of nitrogen-fixing soil bacteria. In practical terms, this meant that certain soils capable of only barely supporting the growth of nitrogen-fixing legume crops could see dramatic improvements from the addition of molybdenum in quantities so small they would be almost impossible to measure without laboratory equipment, sometimes just a few grams of a molybdenum-containing compound spread across an entire field. The discovery carried real weight beyond agronomy circles, because molybdenum-limited nitrogen fixation constrains not just crop growth but the broader capacity of tropical ecosystems to absorb atmospheric carbon dioxide through plant growth, tying a trace soil mineral directly into the planet's carbon cycle in ways researchers are still working to fully quantify.

There is something almost poetic in this finding. An element present in soil at concentrations low enough to be genuinely difficult to detect turns out to be, in certain ecosystems, the single factor standing between a forest that grows vigorously and one that struggles, or between a legume crop that thrives without added fertilizer and one that needs expensive nitrogen inputs trucked in from elsewhere. The tyranny of the tiny, playing out not in a laboratory or an industrial reactor but across entire landscapes.

THE OCEAN'S STRANGE RELATIONSHIP WITH MOLYBDENUM

The story of molybdenum's role in the nitrogen cycle takes an additional, genuinely surprising turn when researchers began studying its behavior in seawater rather than soil. Molybdenum is, relative to most trace metals, reasonably abundant in the ocean, present in dissolved form at concentrations that would seem, on paper, more than sufficient to support vigorous nitrogen fixation by marine cyanobacteria, the photosynthetic bacteria responsible for a substantial share of the ocean's biological nitrogen fixation.

And yet studies of marine nitrogen fixation found that molybdenum's biological availability in seawater is puzzlingly lower than its raw concentration would suggest, a discrepancy researchers eventually traced to an unexpected culprit: sulfate, an entirely different and unrelated ion that happens to be far more abundant in seawater than it is in freshwater. Sulfate, it turns out, is stereochemically similar enough to the form molybdenum takes in oxygenated water, called molybdate, that the two ions compete for the same uptake machinery on the surface of marine microorganisms. Sulfate, present in vastly greater quantities, essentially crowds molybdate out, inhibiting its uptake and reducing its effective biological availability even though the raw dissolved concentration remains high.

This is a subtle and almost mischievous piece of marine chemistry, and it illustrates a theme that recurs throughout the study of trace elements: the mere presence of an element, even in seemingly adequate quantities, does not guarantee that biological systems can actually access and use it. Bioavailability, the technical term for how readily an organism can take up and utilize a given nutrient, depends on a web of chemical interactions and competing ions that can make an element's effective scarcity wildly different from its measured abundance. A soil or an ocean can be, in a strict chemical sense, well supplied with molybdenum, and still leave the organisms living within it functionally starved of it, because some other more abundant substance is quietly blocking the door.

FROM MICROBE TO FIELD TO DINNER TABLE

The practical upshot of a century of research into molybdenum and nitrogen fixation reaches, whether people realize it or not, directly onto dinner tables around the world. Legume crops, from the soybeans that form the backbone of much of the world's protein and cooking oil supply to the alfalfa that feeds dairy cattle to the lentils and chickpeas central to countless traditional cuisines, rely on molybdenum-dependent nitrogenase enzymes housed in bacterial partners living in nodules on their roots. When that molybdenum supply is adequate, these crops can fix a substantial share of their own nitrogen needs directly from the atmosphere, reducing or eliminating the need for expensive synthetic nitrogen fertilizer, the very fertilizer whose industrial production, as the next chapter explores, depends on an entirely different trace element working under drastically different conditions.

Agricultural scientists now routinely recommend molybdenum supplementation for legume crops grown in molybdenum-poor soils, often applying it as a seed treatment, coating legume seeds with a molybdenum-containing compound before planting so that the nitrogen-fixing bacterial partnership has access to the trace metal it needs from the very earliest stages of root nodule formation. The quantities involved remain, by the standards of any other agricultural input, almost comically small, sometimes just a few dozen grams of molybdenum compound per hectare of farmland, an amount that would fit in a coffee mug and yet can meaningfully influence the yield of an entire field.

MOLYBDENUM BEYOND THE FIELD

Molybdenum's usefulness is not confined to agriculture and biology, and a brief detour into its industrial applications helps illustrate a pattern that recurs throughout this book: an element discovered to be essential in one domain often turns out, upon closer inspection, to have an entirely separate and unrelated set of applications elsewhere. In metallurgy, molybdenum is prized as an alloying element in high strength steels, valued for its ability to improve hardenability and maintain strength at elevated temperatures, properties that make molybdenum steel alloys useful in applications ranging from aircraft components to oil and gas drilling equipment operating under extreme pressure and heat. Molybdenum disulfide, a distinct chemical compound rather than the pure metal, is separately valued as an industrial lubricant, its layered crystal structure allowing individual sheets to slide across one another with very low friction, a property exploited in everything from bicycle chains to spacecraft components that must operate reliably in the vacuum of space, where conventional oil-based lubricants would simply evaporate away.

This dual identity, essential biological trace nutrient on one hand and industrially significant structural and lubricating material on the other, mirrors a pattern this book will revisit in its discussion of cobalt, another element whose biological essentiality in vitamin B12 sits alongside an entirely separate and largely unrelated set of industrial applications in battery chemistry and high performance alloys. Nature and industry, it turns out, are frequently drawn to the same handful of elements for entirely independent reasons, a coincidence rooted not in any deep connection between biology and engineering but in the simple fact that the same underlying atomic properties, the specific electron configurations that make certain transition metals so chemically versatile, prove useful across a remarkably wide range of otherwise unrelated applications.

It is worth pausing, at the close of this chapter, to sit with the full scope of what has just been described. A single atom of a rare transition metal, embedded at the center of an elaborate cluster of iron and sulfur atoms, held in place by a scaffold of proteins that evolution spent billions of years refining, sits at the beginning of a chain of consequences stretching from bacterial cells too small to see, through the roots of legume plants, into the crops that feed livestock and people, and ultimately into the nitrogen atoms that make up the proteins in human muscle and bone. Nature discovered this trick long before any human chemist thought to look for it. The next chapter turns to what happened when human chemists, faced with a nitrogen crisis nature's own solution could not scale quickly enough to solve, decided to build their own version of the same reaction, using an entirely different trace element and conditions so extreme that nothing resembling a living cell could have survived them.

Chapter 3

The Ghost in the Furnace

By some estimates, nearly half the nitrogen atoms currently present in the tissue of every human being alive today passed, at some point in the recent past, through a single industrial chemical reaction, one that takes place at temperatures hot enough to glow and pressures found naturally only deep underground. If that estimate holds even approximately true, then a substantial fraction of the protein in your muscles, the DNA in your cells, and the amino acids that built you from the moment of conception owes its existence to a reaction discovered in a German laboratory a little over a century ago, and to a trace ingredient buried inside an unglamorous gray metal catalyst that most people have never heard of and would never think to ask about.

This is the Haber-Bosch process, named for the two chemists whose combined work turned a laboratory curiosity into the industrial backbone of global agriculture, and it stands as perhaps the single clearest demonstration in the modern era of what happens when human ingenuity manages to replicate, and then vastly outscale, a trick that biology had already been quietly performing for billions of years.

A PROBLEM THAT LOOKED LIKE IT MIGHT END BADLY

To understand why the Haber-Bosch process mattered so much, it helps to understand the crisis that preceded it, a crisis that, viewed from the vantage point of the early twentieth century, looked genuinely alarming to the scientists paying close attention to it.

Agriculture had always depended on nitrogen, and for most of history the available sources of usable nitrogen were sharply limited: animal manure, crop rotation with nitrogen-fixing legumes, and, from the mid-nineteenth century onward, imported deposits of sodium nitrate mined from arid regions of Chile, along with guano scraped from islands off the Peruvian coast where seabird colonies had been depositing droppings for centuries. These sources, while valuable, were finite, geographically concentrated, and increasingly strained by the demands of a rapidly growing global population. Agricultural chemists of the era began raising uncomfortable questions about what would happen once the more accessible nitrate deposits were exhausted, questions that carried an unmistakable undertone of impending catastrophe. If the world's food supply depended on a handful of mines in one corner of South America, and those mines had a foreseeable expiration date, what exactly was the plan for feeding the next generation?

The atmosphere, of course, offered an almost absurdly abundant alternative source of nitrogen, four fifths of every breath, freely available everywhere on the planet's surface. The problem, as the previous chapter explored in the context of biological nitrogen fixation, was that atmospheric nitrogen is chemically stubborn almost beyond description, its two atoms locked together by a triple bond that resists being broken by anything short of extreme conditions or a very specifically evolved biological catalyst. Turning that abundant but inert gas into something a crop could use was, at the turn of the twentieth century, one of the most sought-after and most frustratingly elusive goals in industrial chemistry.

A CHEMIST, A HIGH-PRESSURE VESSEL, AND A STUBBORN MOLECULE

Fritz Haber, a German chemist working at the Karlsruhe Institute of Technology, took up this challenge in the first decade of the twentieth century, approaching it with a combination of theoretical rigor and dogged experimental persistence. Haber understood, from the thermodynamics of the reaction, that combining nitrogen and hydrogen gas to form ammonia released energy overall, meaning the reaction was thermodynamically favorable. The trouble, exactly as with the boulder and the hill described in the first chapter of this book, was activation energy: getting the reaction to actually proceed at a useful rate required overcoming the formidable barrier posed by nitrogen's triple bond.

Haber's key insight, refined through years of laboratory experimentation, was that extreme pressure and a suitable catalyst, working together, could coax the reaction forward at temperatures that, while still very hot by everyday standards, were low enough to be industrially manageable. By 1909, Haber had demonstrated the reaction on a laboratory scale, using pressures that were, for the chemistry of the era, extraordinarily high, and a catalyst that pointed the way toward the eventual industrial solution.

A RIVALRY THAT SHAPED THE EARLY RESEARCH

Haber's path to his 1909 demonstration was not a solitary or uncontested one. The chemist Walther Nernst, whose own work on chemical thermodynamics would later earn him a Nobel Prize in his own right, had independently investigated the ammonia synthesis reaction and publicly challenged some of Haber's early experimental results and yield calculations, sparking a pointed scientific dispute that played out partly in the pages of the era's chemistry journals. This disagreement, rather than derailing Haber's research, appears to have sharpened it considerably, pushing him and his laboratory collaborator Robert Le Rossignol toward more careful and more rigorous experimental verification of their claimed results, including the design of specialized high pressure laboratory equipment capable of withstanding the extreme conditions the reaction required without failing catastrophically, equipment that would itself prove to be an essential engineering stepping stone toward Carl Bosch's later industrial scale reactor design.

This kind of contentious, competitive scientific environment, in which rival researchers publicly challenge one another's findings and inadvertently spur one another toward more rigorous work, recurs at several points across the history this book has traced, and it serves as a useful corrective to any tendency to imagine scientific progress as a purely cooperative, harmonious affair. Genuine scientific rivalry, however uncomfortable for the individuals involved, has repeatedly proven to be one of the more effective, if unglamorous, engines of increased experimental rigor across the history of chemistry.

Translating this laboratory demonstration into something that could feed a planet, however, was an entirely different order of problem, and it fell to a chemist and engineer at the German chemical company BASF named Carl Bosch. Bosch faced a set of challenges that had essentially never been solved before at industrial scale: designing vessels and equipment capable of safely containing gases at pressures and temperatures that would cause conventional steel to fail catastrophically, and finding a catalyst that could sustain the reaction reliably, efficiently, and for extended periods of continuous industrial operation, not just for the length of a single careful laboratory trial.

THE PROMOTER'S QUIET GENIUS

It is at this point in the story that a chemist named Alwin Mittasch, working within Bosch's team at BASF, enters the picture, and his contribution goes to the very heart of this book's central theme. Mittasch was tasked with the systematic screening of potential catalysts for the ammonia synthesis reaction, and the scale of that screening effort is worth pausing on: Mittasch and his colleagues are reported to have tested thousands of different catalyst formulations, working through an enormous range of metals, alloys, and additive combinations in search of one that could reliably and efficiently catalyze the reaction under industrial conditions.

The catalyst that eventually emerged from this exhaustive search was based on iron, an element of unremarkable cost and enormous natural abundance, a sensible choice for a process that would need to be run at a scale measured in the millions of tons. But iron alone, it turned out, was not enough. The iron catalyst that actually worked, the one that could sustain efficient ammonia production hour after hour, month after month, inside an industrial reactor, required the addition of trace quantities of two other substances: potassium and aluminum oxide, added not as separate ingredients doing their own independent chemistry but as promoters, substances whose entire function was to make the iron itself perform better.

The mechanism behind each promoter, understood in far greater detail by chemists today than it was by Mittasch and his contemporaries, illustrates two genuinely distinct ways a trace additive can transform a catalyst's performance. Potassium, present in the finished catalyst formulation at a small fraction of the total mass, works as what chemists call an electronic promoter. Sitting on the surface of the iron catalyst, potassium atoms donate electron density to the neighboring iron atoms, and this subtle shift in the iron's electronic environment makes it measurably more effective at binding and weakening the nitrogen molecule's triple bond, the single hardest step in the entire reaction sequence. Aluminum oxide, by contrast, plays what chemists call a structural or textural promotional role. It does not meaningfully alter the iron's chemistry so much as it prevents the iron catalyst particles from sintering, a process in which small catalyst particles gradually fuse together into larger clumps under the heat and pressure of continuous operation, reducing the total surface area available for the reaction to occur on. Scattered through the iron in tiny quantities, aluminum oxide physically wedges itself between iron particles, keeping the catalyst's surface area high and its performance stable across years of continuous industrial use.

Remarkably, the essential composition of this catalyst, iron promoted with potassium and aluminum oxide, has changed very little in the more than a century since Mittasch's original screening work. Modern ammonia plants, processing gas at a scale their early twentieth century predecessors could scarcely have imagined, still rely on catalyst formulations that trace their lineage directly back to the exhaustive trial and error conducted at BASF in the years leading up to the First World War.

THE CATALYST NOBODY EXPECTED TO LAST A CENTURY

There is a further detail about the Mittasch screening effort worth dwelling on, because it illustrates just how much genuine uncertainty surrounded the ammonia synthesis project even after the basic feasibility of iron catalysis had been established. Mittasch and his colleagues did not simply test a handful of promising candidates and settle on the best one. Contemporary accounts of the BASF research effort describe a systematic, almost industrial approach to catalyst discovery itself, testing metal after metal, alloy after alloy, and promoter combination after promoter combination against a standardized set of reaction conditions, patiently building up a body of comparative performance data at a scale that had no real precedent in the chemical industry of the era. This approach, treating the search for a catalyst as its own systematic research program rather than a matter of individual chemical intuition or isolated inspiration, is sometimes credited by historians of chemistry as an early ancestor of what would later become known as combinatorial or high throughput screening, a methodology that remains central to pharmaceutical and materials science research more than a century later, now typically conducted with the assistance of robotic laboratory automation and computational modeling that Mittasch's team, working entirely by hand, could scarcely have imagined.

INSIDE THE REACTOR

The industrial process that Bosch and his engineering team eventually built around Mittasch's catalyst operates under conditions that remain genuinely extreme by ordinary standards. Nitrogen and hydrogen gas are compressed together to pressures typically in the range of one hundred fifty to three hundred atmospheres, roughly the pressure found several kilometers beneath the surface of the ocean, and heated to somewhere between four hundred and five hundred degrees Celsius, hot enough to glow a dull orange if visible. Under these conditions, the gas mixture is passed over beds of the iron catalyst, where a portion of the nitrogen and hydrogen combines into ammonia before the mixture is cooled, allowing the ammonia, which condenses at a higher temperature than the unreacted gases, to be separated out as a liquid. The unconverted nitrogen and hydrogen, and there is always a substantial fraction that fails to react on any single pass through the reactor, is recycled back into the system to be given another chance, a loop that repeats continuously as fresh gas is fed into the process.

This basic architecture, high pressure, elevated temperature, an iron catalyst, and a recycling loop to squeeze maximum efficiency out of every molecule of nitrogen and hydrogen fed into the system, has remained fundamentally unchanged since the earliest industrial plants came online in the years before the First World War. The first commercial ammonia plant, built at BASF's Oppau facility and commissioned in 1913, began with a modest initial capacity of thirty metric tons of ammonia per day. Global ammonia production today, running through plants built on the same underlying chemistry, exceeds one hundred fifty million metric tons annually, a scale-up of more than a thousandfold over the intervening century, achieved without any fundamental change to the core catalytic chemistry Mittasch's team worked out through painstaking trial and error.

A WEAPON FIRST, A HARVEST SECOND

The history of the Haber-Bosch process carries an uncomfortable complication that responsible retellings cannot skip past, one that connects this chapter of chemical history directly to the geopolitics of the early twentieth century. Germany, entering the First World War, faced a British naval blockade that cut off its access to the Chilean nitrate deposits it had previously relied on, and nitrate was not just an agricultural input but also a key ingredient in the manufacture of explosives. The Haber-Bosch process, still in its industrial infancy, was pressed rapidly into service producing nitrogen compounds for munitions rather than fertilizer, extending Germany's capacity to wage war well beyond what its blockaded nitrate supply would otherwise have allowed. Fritz Haber himself, beyond his contribution to ammonia synthesis, played a direct and considerably darker role in Germany's wartime chemical programs, personally overseeing the development and deployment of chlorine gas as a chemical weapon, a legacy that sits in stark and troubling contrast to the humanitarian framing his ammonia synthesis work is often given.

A LIFE OF STARK CONTRADICTIONS

The full arc of Haber's life adds further, still darker dimension to this contrast. Haber, born to a Jewish family in Breslau, converted to Christianity as a young man, a decision biographers have generally interpreted as motivated at least in part by the professional advancement such a conversion could offer within the antisemitic constraints of German academic and civic life at the time. His wartime chemical weapons work drew fierce criticism even from within his own family. His first wife, Clara Immerwahr, herself a trained chemist and reportedly deeply opposed to her husband's chemical weapons research on moral grounds, died by suicide in 1915, shortly after Haber's personal oversight of the first large scale chlorine gas attack at the Second Battle of Ypres, a death many biographers have connected, at least in part, to her documented distress over her husband's wartime work, though the full and precise circumstances remain a matter of ongoing historical discussion. Haber, despite his prominent and patriotic service to Germany during the war, was forced to flee the country after the Nazi rise to power in 1933, on account of his Jewish ancestry, dying in exile in Switzerland the following year, a bitterly ironic end for a chemist whose ammonia synthesis work had, by some later estimates, done more than that of any other single scientist to feed the

German population he had once served with such controversial and consequential loyalty.

It was only after the war's end, as Germany faced the dual pressures of food shortages and the wholesale conversion of wartime industrial capacity to civilian uses, that the Haber-Bosch process was redirected toward its now more familiar role. The Oppau plant, and the larger Leuna facility built up during the war years, began converting their ammonia output primarily into ammonium sulfate and ammonium nitrate fertilizers, feeding nitrogen back into the depleted, war-damaged soils of a country struggling to feed itself. The technology was licensed internationally through the 1920s, spreading well beyond Germany's borders as the global demand for intensified agricultural production, driven by a growing population still recovering from the disruptions of the war, created a ready market for synthetic nitrogen fertilizer wherever it could be produced.

FEEDING A PLANET THAT MULTIPLIED BY FIVE

The scale of the Haber-Bosch process's eventual impact on global food production is difficult to overstate, and the numbers, when laid out plainly, tend to startle people encountering them for the first time. The global human population stood at roughly one point six billion around the year 1900, the years just before Haber's initial laboratory breakthrough. That population has since grown past eight billion, a more than fivefold increase within a little over a century, and a substantial share of the agricultural productivity that made feeding this vastly larger population possible traces directly back to the availability of cheap, abundant, synthetic nitrogen fertilizer produced through the Haber-Bosch process.

The geographer Vaclav Smil, whose scholarship on the topic has become something of a standard reference, has described the Haber-Bosch process in unusually stark terms as the detonator of the human population explosion of the twentieth century, and subsequent research estimates that something close to eighty percent of the nitrogen atoms currently present in the tissue of the average human body have, at some point, cycled through the Haber-Bosch process, whether as fertilizer that fed crops that fed livestock, or as fertilizer that fed crops eaten directly. This is a genuinely strange fact to absorb: a substantial fraction of the physical substance of your own body, the actual nitrogen atoms bound up in your proteins and your genetic material, most likely passed at some point through an industrial reactor operating at extreme temperature and pressure over a catalyst made from iron, potassium, and aluminum oxide.

A SIDE EFFECT MEASURED IN RIVERS AND COASTLINES

The same abundance of synthetic nitrogen that averted the famine many agricultural scientists once feared has, in the decades since, introduced an entirely different category of environmental complication, one that researchers did not fully appreciate until long after the Haber-Bosch process had become thoroughly embedded in global agricultural practice. Not all of the nitrogen fertilizer applied to a field is actually taken up by the crop growing in it. A meaningful fraction, varying considerably depending on soil type, weather, and farming practice, washes off the field entirely, carried by rainfall and irrigation runoff into streams, rivers, and eventually the ocean, where it can trigger a cascade of ecological consequences collectively known to environmental scientists as eutrophication.

Excess nitrogen reaching coastal waters acts as a fertilizer there as well, stimulating explosive growth of algae, and when those algae eventually die and decompose, the decomposition process consumes dissolved oxygen from the surrounding water at a rate that can leave the water unable to support fish, shellfish, or most other forms of marine life, producing what researchers call a dead zone. The Gulf of Mexico has, in recent decades, developed one of the largest recurring dead zones on the planet, driven substantially by nitrogen and phosphorus runoff carried down the Mississippi River from the vast agricultural landscape of the American Midwest, a landscape whose remarkable productivity, as this chapter has described, owes so much to the same Haber-Bosch nitrogen that is now implicated in the degradation of the waters far downstream from where it was originally applied.

This complication does not undo the genuine historical achievement of averting the nitrogen-driven famine that agricultural scientists of the early twentieth century had feared. It does, however, illustrate a pattern that will recur at points throughout the remainder of this book: a trace element intervention, however successful at solving the specific problem it was designed to address, rarely operates in isolation from the broader systems it becomes embedded within, and the same chemistry that feeds a hungry planet can, when deployed without sufficient attention to its downstream consequences, create entirely new problems that take their own additional decades of research and policy attention to fully understand and address.

THE ONGOING SEARCH FOR SOMETHING BETTER

More than a century after Mittasch's original screening work, chemists have not stopped searching for improvements on the iron-based catalyst that made the

Haber-Bosch process possible, and the search itself illustrates just how difficult it remains to displace a well-optimized trace element promoter once it has been established. Researchers have explored alternative catalyst systems, including iron-ruthenium alloys, which show meaningful catalytic activity at somewhat lower temperatures than the traditional iron catalyst, and combinations of cobalt and molybdenum arranged in specific crystal structures with nitrogen atoms incorporated directly into the lattice, an approach that borrows conceptually from the same transition metal chemistry that makes molybdenum so effective in biological nitrogenase.

The motivation behind this continuing research is not merely academic curiosity. The Haber-Bosch process, for all its historic achievement in averting famine, is also enormously energy intensive, consuming, by some estimates, roughly one to two percent of global energy production, largely because of the extreme temperatures and pressures the current iron catalyst requires to function efficiently. A catalyst capable of achieving comparable ammonia yields at meaningfully lower temperature and pressure could reduce the energy footprint of a process that, a century after its invention, remains as essential to feeding the world as it has ever been, and the search for that catalyst continues in laboratories around the world, chasing the same fundamental question Mittasch confronted with his thousands of trial formulations: what trace addition, present in vanishingly small quantity, might unlock a better path over the hill.

It is worth sitting, for a moment, with the strange symmetry between this chapter and the one before it. Nature solved the problem of atmospheric nitrogen fixation using molybdenum, embedded in an elaborate protein scaffold, operating at the gentle temperatures and pressures of ordinary biological life, a solution refined over billions of years of evolutionary trial and error. Human industry solved the same fundamental problem using iron, promoted with potassium and aluminum oxide, operating under conditions of heat and pressure that no living cell could survive, a solution arrived at through a few intense years of deliberate, systematic experimentation. Two entirely different trace elements, two entirely different sets of operating conditions, converging on the same essential outcome: breaking nitrogen's stubborn triple bond and putting the world's most abundant but least accessible element to work feeding the organisms that depend on it. The next chapter turns from the fields these reactions ultimately feed to the human bloodstream, and to a third and very different trace element, one whose absence, rather than its presence, was for decades a mystery that quietly killed the people suffering from it before anyone understood why.

Chapter 4

The Most Critical Speck

"Eat more liver."

For a disease that would eventually be traced to one of the most structurally elaborate molecules in human biochemistry, the first effective treatment could hardly have sounded less sophisticated. And yet, for the patients receiving it in the early decades of the twentieth century, those three words represented something close to a miracle, the difference between a slow, inexorable decline toward death and the real possibility of a full recovery.

The disease in question was pernicious anemia, first documented in the medical literature in 1824 and, for the following century, a diagnosis that carried an almost uniformly grim prognosis. Patients arrived at hospitals pale, exhausted, their red blood cells malformed and unable to carry oxygen effectively, their nervous systems slowly deteriorating in ways that produced tingling, numbness, difficulty walking, and eventually more severe neurological damage. Without treatment, the disease was, true to its name, pernicious in the most literal sense, a slow and often fatal decline that medicine of the era had no reliable way to halt.

The path from that grim diagnosis to a full understanding of its cause runs through one of the more remarkable detective stories in twentieth century medicine, and at the very center of that story sits a single atom of an element that most people, even now, would struggle to name a use for beyond magnets and blue pigment: cobalt.

A DIET THAT SHOULDN'T HAVE WORKED, BUT DID

The first genuine breakthrough in treating pernicious anemia came not from a laboratory but from a carefully designed clinical intervention, conducted by the American physicians George Whipple, George Minot, and William Murphy in the early 1920s. Building on Whipple's earlier observations about the effects of diet on anemia in dogs, the three researchers designed a trial in which pernicious anemia patients were placed on a diet containing large quantities of liver, a food that, on the medical understanding of the time, had no obvious reason to help with a blood disorder whose cause remained entirely unknown.

The results, when they came in, were dramatic and consistent. Patients on the liver-heavy diet showed prompt and substantial remission of their anemia, their symptoms improving in ways that had simply not been achievable through any previously attempted treatment. The finding was significant enough that Whipple, Minot, and Murphy shared the Nobel Prize in Physiology or Medicine in 1934, an unusually rapid recognition for work whose underlying mechanism nobody involved could yet actually explain. Liver clearly contained something, some as yet unidentified factor, capable of reversing a disease that had previously been considered essentially untreatable. What that something was, and why it worked, remained a genuine mystery, one that would take the chemistry world several more decades to fully unravel.

A crucial piece of the puzzle arrived later in the 1920s, when the physician William Bosworth Castle identified a second half of the mechanism. Castle discovered that the human stomach produces a substance he called intrinsic factor, a protein whose job is to bind to the still-unidentified anti-anemia compound found in liver and enable its absorption in the small intestine. Patients suffering from pernicious anemia, it turned out, were often unable to produce sufficient intrinsic factor, meaning that even if their diet contained adequate quantities of the liver-derived compound, their bodies simply could not absorb it. This explained why the disease could develop even in patients whose diets were not obviously deficient in anything: the problem was not a lack of the nutrient in food but a specific failure of the digestive system's ability to extract and absorb it.

ISOLATING THE INVISIBLE INGREDIENT

The identity of the actual compound responsible for curing pernicious anemia remained elusive for another two decades after Castle's discovery of intrinsic factor, a testament to just how minute the effective quantities involved actually were. It was not until 1948 that two separate research teams, working independently and largely unaware of each other's progress, succeeded in isolating and crystallizing the compound. Karl Folkers, working at the pharmaceutical company Merck, and a parallel team at the British pharmaceutical company Glaxo achieved essentially simultaneous breakthroughs, isolating a striking red crystalline compound from liver extracts that, when administered to patients even in extraordinarily small doses, reliably reversed the symptoms of pernicious anemia.

The compound was named vitamin B12, and the sheer potency of the tiny quantities involved surprised researchers accustomed to thinking about vitamins in terms of milligram doses. Patients could be effectively treated with micrograms of the substance, quantities so small they pushed the analytical chemistry of the era to its limits. This was, in itself, an early clue that something unusual was going on at the molecular level, a hint that the compound's effectiveness had less to do with bulk chemical action and more to do with a highly specific, highly efficient interaction with some equally specific biological machinery inside the human body.

What researchers did not yet know, in 1948, was what vitamin B12 actually looked like at the atomic level, how its constituent atoms were arranged into whatever three dimensional structure gave it its remarkable biological potency. That question would occupy one of the twentieth century's most accomplished crystallographers for the better part of a decade.

DOROTHY HODGKIN AND EIGHT YEARS OF STUBBORN CRYSTALS

Dorothy Hodgkin, a British chemist working at Oxford, had already established herself as one of the pioneering figures in the use of X-ray crystallography, a technique that determines the atomic structure of a molecule by analyzing the pattern created when X-rays are diffracted through a crystal of that molecule. Hodgkin had previously used the technique to help unravel the structure of penicillin, work that had proven essential to understanding and eventually synthesizing derivatives of the then-revolutionary antibiotic.

When small red crystals of the newly isolated vitamin B12 became available, grown by the British chemist Lester Smith at Glaxo, Hodgkin turned her considerable expertise toward the problem of determining their structure, beginning work around 1950. What she encountered was a molecule of daunting complexity, the most structurally intricate of all the known vitamins by a considerable margin, and a molecule whose atoms, apart from a lone metal center that stood out clearly in the early diffraction data, remained stubbornly difficult to place with confidence.

That lone metal center turned out to be cobalt, sitting at the structural and functional heart of the entire molecule, coordinated within a complex ring structure called a corrin ring that bears some resemblance to, but is chemically distinct from, the porphyrin ring found at the center of the hemoglobin molecule responsible for carrying oxygen in red blood cells. The presence of that single cobalt atom would eventually explain both the remarkable potency of vitamin B12 in tiny doses and the reason why cobalt itself, an element rarely discussed in the context of human nutrition before this discovery, turned out to be an essential trace mineral for human health, required in the diet specifically because it is the indispensable metal center of this one critical vitamin.

Hodgkin's work on the molecule's full structure was not completed quickly. The scale and complexity of the calculations required, performed in an era before the widespread availability of electronic computers made this kind of crystallographic analysis dramatically faster, meant that her research stretched across roughly eight years of painstaking effort, a period marked by mathematical calculations that filled notebook after notebook and X-ray photographs that, for long stretches, revealed frustratingly incomplete patterns rather than clear answers. When Hodgkin finally announced the complete structure at a Royal Society meeting in March of 1956, having mapped every single atom in one of biology's most structurally complex nonprotein molecules, the achievement was described by contemporaries in terms usually reserved for engineering marvels rather than chemistry, with some comparing its significance to breaking the sound barrier. Hodgkin was awarded the Nobel Prize in Chemistry in 1964, honored specifically for her use of X-ray techniques to determine the structures of biochemical compounds essential to understanding and treating pernicious anemia, alongside her broader body of crystallographic work.

WHY THE BODY NEEDS SO LITTLE, AND WHY THAT LITTLE IS SO HARD TO GET

The structural complexity Hodgkin's work revealed helps explain a fact that had puzzled researchers since the earliest treatment trials: why the human body requires vitamin B12 in such minuscule quantities compared to most other essential nutrients, typically just a few micrograms per day, a dose so small it would be effectively invisible without precise laboratory measurement.

The answer lies in the specific biochemical role the cobalt-centered molecule plays inside human cells. Vitamin B12 functions as an essential cofactor for a small number of enzymes involved in critical cellular processes, including the synthesis of DNA and the maintenance of the protective sheath around nerve cells. Because these enzymatic reactions occur at the molecular level and each molecule of the vitamin can participate in a substantial number of catalytic cycles before needing to be replenished, the body's overall requirement remains extraordinarily small relative to nutrients used in a more structural or bulk capacity.

A RESERVOIR THAT CAN LAST FOR YEARS

One further detail about vitamin B12's physiology helps explain why deficiency, when it does eventually develop, often takes an unusually long time to become clinically apparent, a fact that puzzled early physicians attempting to understand the disease's slow, insidious progression. The human liver maintains a substantial storage reservoir of vitamin B12, sufficient by most clinical estimates to supply the body's needs for several years even in the complete absence of any further dietary intake or absorption. This storage capacity means that a person who develops impaired intrinsic factor production, whether from autoimmune disease or gastrointestinal surgery, may not show any obvious clinical symptoms for a considerable stretch of time, as the body steadily draws down its accumulated hepatic reserve before deficiency symptoms finally begin to emerge. This slow depletion timeline goes a considerable way toward explaining why pernicious anemia, in the decades before its cause was understood, presented clinically as a mysterious, gradually progressive illness rather than one with any clear and immediate dietary trigger a physician might have readily identified through ordinary clinical observation alone. It also means that, unlike some other trace nutrient deficiencies whose symptoms emerge within weeks of an inadequate diet, correcting a B12 absorption problem after it has finally become clinically apparent typically requires sustained treatment over a period of months to fully restore the body's depleted neurological and hematological function, since the underlying nerve damage that severe, prolonged deficiency can cause is not always fully reversible even once adequate vitamin B12 levels have been restored.

The flip side of this efficiency is that vitamin B12 deficiency, when it occurs, can be surprisingly difficult to diagnose and treat, precisely because the quantities involved are so small and the absorption pathway, dependent as it is on Castle's intrinsic factor, is so specific and so easily disrupted. Patients with autoimmune conditions affecting the stomach's ability to produce intrinsic factor, patients who have undergone certain types of gastrointestinal surgery, and, notably, individuals following strict vegan diets without appropriate supplementation, since vitamin B12 is found reliably only in animal-derived foods, all face meaningfully elevated risk of deficiency, a risk that modern medicine addresses through injections or high-dose oral supplements that bypass or overwhelm the limitations of the natural intrinsic factor absorption pathway.

COBALT'S QUIET DOUBLE LIFE

There is something worth noting about cobalt's broader relationship to human industry and human biology, a duality that recurs, in various forms, throughout this book. The same element sitting at the functional heart of a vitamin essential to human nervous system health is also, in an entirely unrelated context, a critical industrial metal, used in the manufacture of high-strength alloys for jet engine turbine blades, in powerful magnets, and, as a later chapter of this book explores in detail, in the cathode materials of many modern lithium-ion batteries. The demand for cobalt in battery manufacturing has, in recent years, drawn considerable attention to the mining conditions under which much of the world's cobalt supply is extracted, particularly in the Democratic Republic of the Congo, which accounts for a substantial majority of global cobalt production and where labor and environmental conditions in parts of the mining sector have drawn sustained scrutiny from human rights organizations and journalists.

This double life, essential trace nutrient on one hand, contested industrial commodity on the other, is a reminder that the story of trace elements told in this book is never purely a story of elegant chemistry unfolding in isolation. It is also, inescapably, a story about the human systems, economic and political, that determine how these elements are extracted from the earth, who profits from that extraction, and under what conditions the people doing the physical work of mining actually labor. Cobalt's role in curing pernicious anemia and its role in powering a smartphone battery are chemically unrelated, but both draw on the same finite global supply of an element that, despite its outsized importance across two entirely different domains of modern life, remains present in the earth's crust at concentrations low enough to make it, by any reasonable definition, genuinely scarce.

A MOLECULE STILL WORTH STUDYING

The story of vitamin B12 did not end with Hodgkin's structural determination in 1956. The molecule's remarkable complexity, once fully mapped, presented organic chemists with an almost irresistible challenge: could such an intricate structure actually be synthesized from scratch in a laboratory, without relying on extraction from natural sources? The answer came from a collaboration between the chemist Robert Woodward at Harvard and Albert Eschenmoser at the Swiss Federal Institute of Technology, whose combined research groups spent roughly eleven additional years after Hodgkin's structural work developing a complete synthetic pathway for vitamin B12, a project that ultimately required more than ninety separate chemical reactions carried out by over a hundred collaborating researchers across the two institutions. The stereochemical puzzles this synthesis project encountered along the way proved significant enough in their own right to give rise to a set of general chemical principles, now known as the Woodward-Hoffmann rules, that remain a standard part of organic chemistry education today.

Taken together, the effort to understand, isolate, structurally determine, and finally synthesize vitamin B12 produced, across its various stages, three separate Nobel Prizes in Chemistry and one in Physiology or Medicine, an extraordinary concentration of scientific recognition for a single molecule, and a molecule whose functional core, the part actually responsible for all this biochemical and industrial significance, consists of one atom of an element most people could not confidently place on the periodic table.

THE WIDER FAMILY OF METAL COFACTORS

Vitamin B12 is, for all its structural drama, only one member of a much larger family of biological molecules that rely on a trace metal center to perform their function, and placing it within that broader family helps clarify why this particular pattern, a single metal atom sitting at the heart of an otherwise elaborate protein or organic scaffold, recurs so often across biochemistry. Hemoglobin, the oxygen carrying protein responsible for the red color of blood, relies on iron held within a ring structure closely related to the corrin ring at the heart of vitamin B12, iron whose particular electronic properties allow it to bind and release oxygen molecules reversibly as blood circulates between the lungs and the body's tissues. Chlorophyll, the pigment that allows plants to capture light energy for photosynthesis, relies on an almost structurally identical ring system, but with magnesium substituted at its center rather than iron or cobalt, a striking illustration of how evolution has repeatedly reused the same basic structural architecture across radically different biological functions, simply by swapping in a different trace metal at the functional core.

This recurring pattern, an elaborate organic scaffold whose function depends entirely on a single, precisely positioned metal atom, suggests something important about how biological catalysis evolved in the first place. Rather than inventing an entirely new solution for every metabolic challenge an organism might face, evolution appears to have favored a strategy of reusing and adapting a comparatively small set of proven structural templates, tuning their function by substituting different trace metals into an otherwise similar scaffold, an economical evolutionary strategy that echoes, in its own biological register, the industrial chemist's strategy of taking a proven catalyst architecture and adjusting its trace promoter composition to suit a new application, the very strategy Alwin Mittasch pursued in developing the Haber-Bosch catalyst described in this book's previous chapter.

It is a fitting close to this chapter to note the scale involved one final time. A single cobalt atom, present in a human body at a total quantity measured in milligrams, sitting at the heart of a molecule required in daily doses too small to see without magnification, stands between ordinary human health and a disease that, for the first hundred years after its clinical description, had no reliable cure at all. The next chapter turns to a different trace element, one whose story runs in something closer to the opposite direction: not a molecule whose absence caused a rare and puzzling disease, but a mineral whose absence, across entire regions and for generations, produced a public health crisis so widespread and so visible that it left a permanent mark on the medical geography of the modern world.

Chapter 5

The Element That Counts Silently

At the time of the American draft for the First World War, roughly a third of the young men called up for service from one particular region of Michigan were rejected as physically unfit, and the reason had nothing to do with weak hearts, poor eyesight, or any of the more commonly imagined disqualifying conditions. Their necks were too thick to fit the standard military uniform collar. The cause was goiter, a visible swelling of the thyroid gland, and it was so common across parts of the upper Midwest that public health surveys conducted in the 1920s found goiter rates among schoolchildren in portions of Michigan, Minnesota, and Wisconsin running as high as seventy to one hundred percent.

This was not, as some contemporaries speculated, an infection sweeping through the region, nor was it linked to poor sunlight or bad air, despite both explanations having genuine advocates among the roughly forty two competing theories about the cause of endemic goiter catalogued in an 1867 medical publication. The actual cause was both simpler and, in retrospect, almost unbearably easy to fix: the soil across this swath of the country, along with the soil of the Pacific Northwest and the Appalachian region, was naturally poor in a single trace element, iodine, and in an era when most food was grown and consumed locally, an iodine-poor soil meant an iodine-poor diet for the entire human and livestock population living on it.

These iodine-starved regions became known, in the grim shorthand of public health researchers at the time, as the goiter belt, and understanding how that belt came to be, and how it was eventually and almost completely eliminated within the span of a single generation, is the subject of this chapter.

A VIOLET VAPOR AND AN ACCIDENTAL DISCOVERY

The story of iodine's medical significance begins, somewhat improbably, with gunpowder. In 1811, the French chemist Bernard Courtois was working to improve the efficiency of extracting saltpeter from seaweed ash, a process relevant to gunpowder manufacture, when he noticed a striking violet vapor rising from one of his vats during an experiment involving concentrated sulfuric acid. That vapor, upon cooling, condensed into dark crystals of a previously unidentified element, one that would eventually be named iodine, drawing on the Greek word for the color violet in recognition of its distinctive vapor.

Courtois had, without intending to, uncovered an element that would go on to transform global nutrition, though it would take another century before that connection to human health became fully understood. The thyroid gland, a butterfly-shaped organ sitting at the base of the neck, requires iodine as an essential raw material for synthesizing thyroid hormones, chemical messengers that regulate metabolism, growth, and development throughout the body. When dietary iodine is insufficient, the thyroid gland, in a physiological attempt to compensate, enlarges in size, working harder to extract every available trace of the mineral from a bloodstream that simply does not contain enough of it. This enlargement is the visible swelling known as goiter, and it represents, in a sense, the thyroid gland's desperate and ultimately futile attempt to solve a supply problem through sheer effort rather than raw material.

Goiter, unpleasant and disfiguring as it could be, was in many respects the least severe consequence of iodine deficiency. Far more serious was the effect of iodine deficiency during pregnancy, when an insufficient maternal supply of thyroid hormone could produce irreversible damage to fetal brain development, a condition in its most severe form known historically as cretinism, associated with profound and permanent intellectual disability. Iodine deficiency during pregnancy and early childhood, researchers eventually established, represented one of the world's leading preventable causes of impaired cognitive development, a public health burden that, because its effects were gradual, widespread, and not always dramatically visible in the way goiter was, took considerably longer to be fully appreciated than the more obvious physical symptom of a swollen neck.

A CLINICAL TRIAL IN AKRON

The scientific breakthrough that eventually connected iodine supplementation to goiter prevention in humans came from a carefully designed clinical trial conducted between 1917 and 1920 in Akron, Ohio, led by the pathologist David Marine and a medical student named Oliver Kimball. Marine had already conducted earlier research suggesting a link between iodine and thyroid function, and the Akron trial was designed to test, rigorously and at meaningful scale, whether iodine supplementation could actually prevent the development of goiter in a population known to be at elevated risk.

The trial involved a substantial cohort of over two thousand schoolchildren, divided between a group receiving iodine supplementation and a control group that did not. The results provided clear and compelling evidence that iodine supplementation dramatically reduced the incidence of goiter, findings robust enough to catch the attention of public health advocates looking for a scalable intervention that could be applied not just to a research cohort but to entire populations living in iodine-deficient regions.

Marine's own path to this research carried a personal dimension that adds useful context to his persistence. He had begun investigating goiter's relationship to iodine years earlier through observational studies of both animal populations and human patients in the greater Cleveland area, accumulating a body of preliminary evidence well before the Akron trial's formal launch, at a time when the prevailing medical opinion regarding goiter's cause remained genuinely unsettled and Marine's iodine hypothesis was merely one of several competing explanations still being actively debated within the medical community. The scale and rigor of the eventual Akron trial, involving careful record keeping across a cohort of schoolchildren large enough to yield statistically convincing results, represented a considerable advance in methodological sophistication over the smaller, less systematic observational studies that had preceded it, and it is this methodological rigor, as much as the underlying iodine hypothesis itself, that historians of medicine credit with finally moving the scientific consensus decisively in Marine and Kimball's favor.

Switzerland, wrestling with its own severe iodine deficiency problem rooted in the iodine-poor soils of its mountainous terrain, became one of the first places in the world to implement a population-wide solution, adopting salt iodization in the early 1920s. Salt was, from a public health engineering perspective, an almost perfect vehicle for delivering a trace nutrient to an entire population: it was consumed by virtually everyone regardless of income or dietary preference, in reasonably predictable quantities, and it could be fortified with a small addition of a stable iodine compound at the point of production without requiring any change in consumer behavior or awareness.

MICHIGAN, MAY 1, 1924

In the United States, the push to replicate Switzerland's success fell largely to a University of Michigan pediatrician named David Cowie, who argued forcefully that fortifying table salt with iodine could eliminate the goiter epidemic afflicting his state and the broader Midwestern goiter belt entirely. Cowie's advocacy succeeded, and on May 1, 1924, iodized salt appeared for the first time on store shelves in Michigan, marking the practical beginning of one of the more understated and yet more consequential public health interventions of the twentieth century.

The results, once they had time to unfold across a generation of Michigan schoolchildren, were dramatic. Goiter rates in the formerly afflicted regions plunged within a matter of years, and the visible epidemic that had once left roughly a third of local draft candidates unable to fit a standard uniform collar essentially disappeared from the American public health landscape within a generation. The intervention was not without occasional complications; some adults with long-standing nodular goiters experienced episodes of iodine-induced hyperthyroidism as their thyroid function adjusted to a sudden increase in available iodine after years of adaptation to scarcity, a reminder that even a broadly beneficial trace element intervention can carry localized complications for a subset of the population it reaches. But measured against the scale of the problem it addressed, the public health benefit of salt iodization was, by any reasonable accounting, overwhelming.

A CAMPAIGN BUILT ON PERSUASION AS MUCH AS SCIENCE

The scientific case for iodized salt, however compelling in retrospect, did not translate automatically into public acceptance, and the campaign to popularize the new product in Michigan required a sustained effort of public communication that historians of public health have since studied as an early and instructive example of applied science communication. Cowie and his collaborators worked closely with local newspapers, physicians, and school health programs to explain, in terms an ordinary consumer could readily understand, why a product that looked, tasted, and cost essentially the same as ordinary table salt was nonetheless worth specifically seeking out. Some early skepticism from consumers concerned about the safety of adding any unfamiliar substance to their food had to be actively and patiently addressed, a public communication challenge that would recur, in various forms, across many of the other population-level trace element interventions this book describes, including the far more contentious and ultimately much slower public reckoning with the dangers of leaded gasoline explored in a later chapter.

The Michigan iodization campaign's eventual success established a template that other American states, and eventually the federal government's own public health guidance, would draw upon in subsequent decades, and it stands as an instructive early case study in a pattern this book returns to repeatedly: the scientific discovery of a trace element's importance is very often only the first, and not necessarily the hardest, step in actually translating that discovery into a durable, population-wide public health benefit. The second step, persuading an often skeptical public and a frequently under-resourced public health infrastructure to actually adopt and sustain the intervention over years and decades, has proven, across the many trace element stories this book tells, to be at least as demanding as the original scientific discovery itself.

THE IQ POINTS NOBODY NOTICED WERE MISSING

Perhaps the most striking piece of evidence for the true scale of iodine deficiency's historical impact comes not from goiter statistics but from a very different kind of data source: aptitude test results administered to American military recruits. Economists studying these historical records have estimated that, for the roughly one quarter of the American population that had been most severely iodine deficient prior to widespread salt iodization, the introduction of iodized salt raised average population IQ by a striking fifteen points, a magnitude of cognitive improvement that would be considered extraordinary if achieved through any modern educational or nutritional intervention, and one that occurred almost invisibly, distributed across an entire generation, achieved through nothing more dramatic than a pinch of iodine stirred into salt at a concentration too small for anyone eating it to notice by taste, smell, or appearance.

This finding carries a sobering implication that extends well beyond the specific historical case of the American Midwest. If a trace element deficiency affecting roughly a quarter of a population could suppress average cognitive development by fifteen IQ points without producing any single dramatic, attention-grabbing event, a war, an epidemic, a famine, capable of forcing the problem onto the public agenda, how many similarly invisible nutritional deficiencies might still be quietly constraining human potential in populations around the world today, absent the kind of sustained scientific attention and public health infrastructure that eventually identified and solved the iodine problem in wealthier nations?

A GLOBAL CAMPAIGN, STILL UNFINISHED

The success of salt iodization in the United States and Switzerland eventually inspired a broader international public health movement, though its global adoption proceeded considerably more slowly and unevenly than its early champions might have hoped. By the early 1990s, the World Health Organization estimated that more than two billion people worldwide, roughly a third of the global population at the time, remained at risk of iodine deficiency, a staggering figure given how straightforward and low-cost the solution had already proven to be decades earlier in the regions that had implemented it.

The years since 1993, when the World Health Assembly formally endorsed universal salt iodization as a global public health strategy, have seen substantial, if uneven, progress. India, drawing on a sustained national commitment, achieved universal salt iodization coverage exceeding ninety percent across urban areas of its Central, North, and North-East zones by 2015. Ethiopia's national coverage of iodized salt climbed from just over four percent in 2005 to ninety five percent by 2014, an extraordinarily rapid public health transformation achieved within a single decade. More than one hundred twenty countries now maintain some form of mandatory salt iodization program, and the global percentage of households consuming adequately iodized salt has risen from roughly one quarter in 1990 to approximately two thirds today, a genuine and significant public health achievement, even as researchers continue to document an estimated two billion people worldwide who remain iodine deficient, particularly in regions where iodized salt remains scarce, inconsistently available, or prohibitively expensive.

WHAT MADE SOME NATIONAL CAMPAIGNS SUCCEED WHERE OTHERS STRUGGLED

Comparative public health research examining why some countries achieved rapid, near universal salt iodization coverage while others have struggled for decades to reach even modest coverage levels has identified several recurring factors that appear to distinguish the more successful national campaigns. Sustained, high level political commitment, of the kind Ethiopia's health ministry maintained through its decade long push toward ninety five percent coverage, has consistently proven more important than any single technical or economic factor in explaining program success, since salt iodization requires ongoing coordination across salt producers, distributors, retailers, and public health monitoring agencies that only sustained institutional attention can reliably maintain over the years a national rollout typically requires. Countries that concentrated salt production among a comparatively small number of large scale producers, easier to regulate and monitor than a highly fragmented small scale salt industry, have also generally achieved more rapid and more consistent coverage than countries where salt production remained dispersed across thousands of small, difficult to monitor local producers, a structural factor largely outside the direct control of the public health officials running any given national iodization campaign.

The unevenness of this progress carries its own lessons about the practical challenges of trace element public health interventions, challenges that go well beyond simply identifying the correct nutrient and the correct dose. Some countries have found that monitoring and enforcement of salt iodization laws proved far more difficult in practice than the initial legislation suggested, with unenforced mandates and absent quality monitoring systems undermining the intended benefit in Latin American countries during the mid twentieth century, for example, despite the passage of iodization laws in the 1950s and 1960s. Even in regions with well-established programs, researchers have documented significant variability in the actual iodine content of salt reaching consumers, with some studies finding samples ranging from inadequately iodized to substantially over-iodized within the same distribution system, underscoring that a trace element intervention, precisely because the effective dose is so small, requires genuinely careful and sustained quality control to work as intended across an entire population and supply chain.

WHY IODINE STILL MATTERS IN A WORLD OF ABUNDANT SALT

It would be reasonable to assume that, in wealthier nations with robust and long-established food fortification programs, iodine deficiency has become a solved problem, a historical curiosity rather than an ongoing public health concern. The reality, as endocrinologists studying the issue have continued to document, is somewhat more complicated. Dietary patterns have shifted considerably since the early and mid twentieth century, with a growing share of salt consumption in many countries now coming from processed and restaurant food, which frequently uses non-iodized salt, rather than from iodized salt purchased and used directly in home cooking. Researchers studying iodine status in parts of Europe, including regions of Italy and Switzerland that had achieved substantial historical success with iodization programs, have found evidence of renewed mild iodine deficiency emerging in more recent survey data, a reminder that the public health gains achieved through trace element fortification are not necessarily permanent once achieved, but require ongoing attention as dietary habits and food supply chains continue to evolve.

THE OTHER SIDE OF THE LEDGER

Iodine's story, like several others in this book, carries an important caveat that resists a purely triumphant telling: the relationship between iodine intake and thyroid health is not simply a matter of more being better. Excessive iodine intake, whether from over-fortified salt, high-dose supplements, or, in some documented cases, iodine-rich seaweed consumed in unusually large quantities, can itself trigger thyroid dysfunction, including a form of hyperthyroidism similar to the complication observed among some adults during the initial rollout of iodized salt in the 1920s. Public health researchers monitoring global iodine status have documented that a small but consistent share of national iodization programs, particularly in the years immediately following their introduction, have overshot the intended target range, producing populations with iodine intake high enough to raise its own set of clinical concerns rather than resolving the deficiency the program was designed to address.

This dynamic, a narrow therapeutic window in which too little causes one set of problems and too much causes a different set of problems, is not unique to iodine among the trace elements this book explores, but iodine's case illustrates it with particular clarity, precisely because the intervention involved, salt fortification, operates at a population level rather than through individualized medical dosing. A physician prescribing vitamin B12 injections, discussed in the previous chapter, can titrate an individual patient's dose based on that patient's specific bloodwork. A national salt iodization program, by contrast, must select a single fortification level intended to serve an entire population's needs simultaneously, a population whose baseline dietary iodine intake will inevitably vary considerably from one individual and one region to the next, depending on local food habits, proximity to iodine-rich seafood, and the iodine content of local soil and water. Getting this population-level dosing right, precisely enough to correct deficiency in the neediest segment of the population without over-correcting in segments that were already receiving adequate iodine from other dietary sources, has proven to be a genuinely difficult and ongoing public health engineering challenge, one that national salt iodization programs continue to refine through periodic monitoring and adjustment even in countries where the program has been running for the better part of a century.

There is a broader lesson embedded in iodine's history that resonates throughout the other chapters of this book, though nowhere else quite so starkly. The trace elements this book explores tend to matter most, and to be understood least, precisely because their effects are so quiet. A catalytic converter's platinum silently transforms exhaust gases without any visible sign of the chemistry occurring inside it. An iron catalyst inside an ammonia plant silently fixes nitrogen at a scale that feeds billions of people who will never think about the reactor that made their meal possible. And a pinch of iodine, stirred into salt at a concentration too small to taste, silently protected the developing brains of an entire generation of children from a form of harm so gradual and so widely distributed that it took economists studying military aptitude test scores, decades later, to fully reveal the scale of what had been quietly prevented.

The next chapter turns to a very different kind of quiet chemistry, one that happens not in soil or in the human thyroid gland but underneath the hood of virtually every gasoline-powered vehicle on the road, where three more trace elements, working together in combinations measured in single-digit grams, silently convert some of the most dangerous gases produced by modern civilization into some of the least dangerous, tens of thousands of times an hour, for the entire working life of the vehicle they are installed in.

Chapter 6

The Invisible Catalyst Box

Somewhere beneath the floor of the car you may have driven today sits a cylindrical metal canister, roughly the size of a large coffee thermos, packed with a ceramic honeycomb structure and coated with a combined mass of precious metal that likely weighs less than a few grams, less than a single coin. That unassuming component has, over the course of the vehicle's lifetime, processed the exhaust from thousands of gallons of burned gasoline, converting gases that would otherwise poison the air of every city on Earth into gases that are, for the most part, harmless. It does this work continuously, silently, without moving parts, without maintenance in the ordinary sense, and without most drivers ever giving it a single thought.

This is the catalytic converter, and its story is, in miniature, a case study in exactly the kind of chemistry this book is devoted to: an enormous environmental and public health problem, solved not through some dramatic new technology but through the extraordinarily precise deployment of a handful of rare metals, present in quantities so small that a typical converter's entire precious metal content could be held in the palm of one hand.

THE PROBLEM RIDING BEHIND EVERY ENGINE

To understand why the catalytic converter became necessary, it helps to understand what an internal combustion engine actually produces when it burns gasoline, and how far that output falls short of the clean combustion the engine's designers might wish for. In a perfect world, an engine burning gasoline, a hydrocarbon fuel, would combine every molecule of fuel completely with oxygen from the air, producing nothing but carbon dioxide and water vapor as exhaust. Real engines, running under the messy and variable conditions of actual driving, achieve nothing close to this ideal. Incomplete combustion leaves behind carbon monoxide, a colorless and highly toxic gas that interferes with the blood's ability to carry oxygen, along with unburned hydrocarbons that contribute to the formation of ground-level smog. Meanwhile, the extreme heat and pressure inside the combustion chamber forces some of the otherwise inert nitrogen in the intake air to react with oxygen, producing a family of compounds called nitrogen oxides, which contribute both to smog formation and to acid rain.

By the 1960s, the cumulative effect of millions of vehicles producing these pollutants had become impossible to ignore in cities across the industrialized world, with Los Angeles serving as a particularly visible and frequently cited example of severe automotive smog. The regulatory response, most notably the Clean Air Act in the United States, demanded radical cuts in vehicle emissions, cuts that engine design improvements alone could not realistically achieve. What was needed was a way to treat the exhaust after it left the engine, cleaning up the mess of pollutants before they were released into the atmosphere, and the technology that emerged to meet this demand became standard equipment on American vehicles beginning in 1975, with similar regulatory mandates following in Europe and other regions through the 1990s.

A TRIO OF METALS, EACH WITH A DIFFERENT JOB

The catalytic converter that emerged to solve this problem relies on a small family of closely related metals from the platinum group, most centrally platinum, palladium, and rhodium, deployed together in what engineers call a three-way catalyst, so named because it addresses three distinct classes of pollutant simultaneously. Understanding why these particular three metals, and not some more abundant and less expensive alternative, ended up at the heart of automotive emissions control requires understanding both the extreme conditions inside an exhaust system and the very specific chemistry each pollutant requires to be neutralized.

The environment inside a vehicle's exhaust system is genuinely harsh, cycling through wide swings of temperature, exposed to a chemically aggressive mixture of combustion byproducts, and required to function reliably for years without maintenance. Ordinary base metals, researchers exploring alternatives have consistently found, tend to corrode, melt, or simply lose their catalytic activity under these conditions far more quickly than the application demands. Gold and silver, tested as potential candidates in early research, were eliminated because of insufficient durability and catalytic activity for the task. Other platinum group candidates, including ruthenium, iridium, and osmium, showed promising catalytic activity in laboratory testing but were ultimately ruled out because they form volatile oxide compounds at the high temperatures found inside an operating exhaust system, meaning they would gradually and unacceptably evaporate away rather than remaining in place to keep catalyzing reactions.

This narrowing process left platinum and palladium as the primary candidates for the earliest catalytic converters, brought into commercial use to address the oxidation half of the emissions problem: converting carbon monoxide and unburned hydrocarbons into carbon dioxide and water through reaction with oxygen present in the exhaust stream. As regulatory standards tightened further to specifically target nitrogen oxide emissions, a third metal, rhodium, was incorporated into the catalyst formulation, because rhodium proved distinctly and reliably superior to platinum or palladium at catalyzing the opposite kind of reaction, a reduction reaction in which nitrogen oxides are stripped of their oxygen atoms and converted back into harmless nitrogen gas.

The distinction between these two categories of reaction, oxidation and reduction, running in opposite chemical directions, explains why a modern catalytic converter is not simply a single uniform coating of precious metal but a carefully engineered, layered structure. Platinum and palladium, positioned to handle the oxidation of carbon monoxide and hydrocarbons, work most effectively when exhaust gas contains a relative abundance of oxygen. Rhodium, tasked with reducing nitrogen oxides, works best under the opposite condition, when oxygen is comparatively scarce and the surrounding gas mixture is chemically primed to strip oxygen away from other molecules rather than add it. A well-engineered three-way catalyst manages this apparent contradiction by relying on precise, continuous electronic control of the engine's air to fuel ratio, oscillating it rapidly around a chemically balanced point so that both the oxidation and reduction reactions can each find, in alternating moments, the exhaust chemistry they need to proceed efficiently. Rhodium's particular value in this arrangement extends beyond simple effectiveness at reducing nitrogen oxides. It also produces meaningfully less ammonia as an unwanted side product of that reduction reaction than platinum or palladium would if used for the same purpose, and ammonia produced inside the catalytic converter carries the unwelcome risk of being reconverted back into nitrogen oxide pollutants further downstream, undermining the very purpose the converter was designed to achieve.

A FOOTBALL FIELD FOLDED INTO A COFFEE THERMOS

The engineering challenge of making this three-metal system commercially viable centered on a problem already introduced in this book's opening chapter: precious metals, by definition, are expensive, and a catalytic converter containing a solid or thickly plated mass of platinum, palladium, and rhodium would add a prohibitive amount of cost to every vehicle it was installed in. The solution, refined over decades of automotive engineering, involves depositing these metals as an extraordinarily thin coating, sometimes just a few micrometers thick, across the surface of a ceramic support structure shaped into a dense honeycomb of parallel channels.

This honeycomb geometry accomplishes something that might otherwise seem like a contradiction: it packs an enormous amount of surface area into a remarkably compact physical volume. A typical catalytic converter's honeycomb structure, if its internal surface were somehow unfolded and laid completely flat, would cover an area comparable to a football field, all contained within a canister that fits comfortably beneath the floor of an ordinary passenger car.

THE CERAMIC BENEATH THE METAL

The honeycomb substrate itself deserves a moment of attention, since it represents a separate and largely uncelebrated materials science achievement underlying the catalytic converter's success. Most modern converter substrates are manufactured from cordierite, a magnesium aluminum silicate ceramic prized for an unusual combination of properties that make it particularly well suited to this specific application: a very low coefficient of thermal expansion, meaning it does not significantly expand or contract as it repeatedly heats and cools through the wide temperature swings of ordinary engine operation, and sufficient mechanical strength to survive the vibration and thermal shock of automotive use across a vehicle's operational lifetime without cracking or crumbling. Engineers manufacture this ceramic substrate with wall thicknesses measured in fractions of a millimeter and channel densities that can exceed several hundred individual passages per square inch of cross section, a manufacturing precision that, much like the precious metal coating it supports, represents decades of incremental engineering refinement rather than a single inventive breakthrough.

Exhaust gas flowing through the converter passes across this vast folded surface, coming into contact with the thin precious metal coating at millions of points simultaneously, which allows even a very small total mass of platinum, palladium, and rhodium, often just a handful of grams combined, to process an enormous volume of exhaust gas continuously over the vehicle's operational lifetime.

Beneath the immediate precious metal coating, engineers layer in additional supporting materials that play their own promotional and structural roles, echoing the promoter chemistry described in the previous chapter's account of the Haber-Bosch catalyst. Cerium oxide and zirconium oxide, incorporated into the catalyst's washcoat, serve as oxygen storage compounds, absorbing and releasing oxygen as the engine's air to fuel ratio fluctuates in real time, helping to smooth out the oscillations the three-way catalyst depends on and improving the stability and durability of the precious metal catalysts sitting above them. None of these supporting materials perform the core catalytic chemistry themselves, but their presence, again in comparatively small quantities, measurably improves the performance and longevity of the precious metals that do.

THE ECONOMICS OF SCARCITY, MEASURED IN GRAMS

The reliance of catalytic converter technology on platinum, palladium, and rhodium has created a genuinely unusual commodity market, one shaped almost entirely by a handful of industrial applications rather than the broader jewelry and investment demand that drives markets for gold and silver. Rhodium in particular has proven to be an extraordinarily scarce and price-volatile metal, its market price at various points climbing to levels many multiples higher than gold, driven by the combination of its irreplaceable role in nitrogen oxide reduction and its genuinely limited global supply, concentrated overwhelmingly in mining operations in South Africa.

The volatility this market concentration produces is worth illustrating with a concrete sense of scale. Rhodium prices have, at various points over the past two decades, swung by a factor of ten or more within a span of just a few years, driven by comparatively small shifts in South African mining output, changes in automotive production volumes, or speculative trading activity, a degree of price volatility rarely seen in more broadly traded commodities and one that automotive manufacturers, needing to plan vehicle production and pricing years in advance, have found genuinely difficult to hedge against using the conventional financial instruments available for more liquid, widely traded commodities like gold, oil, or copper. This volatility has itself become a further motivation, alongside the underlying supply security concerns already discussed, for the automotive industry's sustained research investment into rhodium reduced and rhodium free catalyst formulations.

This concentration of supply has created real strategic vulnerability throughout the automotive industry's supply chain, prompting sustained research into catalyst formulations that could reduce reliance on the most scarce of the three metals without sacrificing emissions performance. Researchers have explored layered catalyst architectures that position a protective outer layer of platinum above an inner layer containing smaller quantities of rhodium and palladium, shielding the more vulnerable inner metals from the catalyst poisons, such as lead and phosphorus compounds, historically found in some fuel and engine oil formulations. This layered approach has allowed some catalyst designs to function effectively with rhodium content reduced to concentrations as low as roughly two thousandths of one percent by weight of the total catalyst, an almost absurdly small figure that nonetheless reflects real, carefully engineered chemistry rather than mere cost-cutting, since rhodium below a certain minimum concentration simply stops providing adequate nitrogen oxide reduction performance regardless of how cleverly it is positioned within the catalyst structure.

Parallel research efforts, driven by a combination of cost pressure and supply security concerns, have explored the possibility of three-way catalysts that eliminate rhodium entirely, relying instead on carefully engineered combinations of platinum and palladium alongside elevated concentrations of cerium oxide to achieve comparable nitrogen oxide reduction performance without rhodium's contribution. The very existence of this ongoing research effort, decades after the three-way catalyst's basic architecture was first commercialized, underscores just how tightly automotive engineers remain constrained by the fundamental chemistry involved: there is, even now, no abundant and inexpensive substitute that can fully replicate what these particular platinum group metals accomplish, which is precisely why a market in recovering and recycling precious metals from worn-out catalytic converters has grown into a substantial industry of its own, treating discarded automotive parts as a genuine, if modest, source of newly available platinum, palladium, and rhodium.

A QUIET LEGACY WRITTEN INTO CITY AIR

The cumulative public health impact of catalytic converter technology, deployed across hundreds of millions of vehicles worldwide over the past half century, is difficult to calculate with precision but is generally regarded by air quality researchers as substantial, contributing meaningfully to measurable declines in urban concentrations of carbon monoxide, unburned hydrocarbons, and nitrogen oxides across cities that had, in the middle decades of the twentieth century, struggled with severe and visible smog problems. The technology has not been without its own environmental complications; researchers studying urban environments have documented measurable releases of platinum group metals into roadside dust and soil as catalytic converters gradually wear over the course of their operational lifetime, a subtler and less immediately dangerous concern than the pollutants the converters are designed to eliminate, but one that continues to receive ongoing scientific attention as researchers work to understand its long-term environmental significance.

AN UNEXPECTED CRIME WAVE

The precious metal content of catalytic converters has, in recent years, generated a genuinely unusual and largely unanticipated social consequence, one that illustrates just how much economic value can be concentrated into an object most vehicle owners never think about. Rising prices for platinum, palladium, and especially rhodium through the late 2010s and early 2020s made the metal content of a single catalytic converter, sitting unattended beneath a parked vehicle, valuable enough to attract organized theft on a scale that surprised even veteran law enforcement officials. Thieves, working with a car jack and a portable reciprocating saw, can remove a catalytic converter from beneath a parked vehicle in a matter of minutes, and the resale value of the extracted precious metals, sold to scrap metal dealers who in turn sell the material into legitimate industrial recycling channels, proved lucrative enough that catalytic converter theft rates climbed sharply across a number of countries during this period, prompting a range of responses including vehicle owners etching identifying numbers into their converters, manufacturers redesigning mounting hardware to be more theft resistant, and legislators in several jurisdictions introducing new regulations governing how scrap metal dealers document the provenance of the converters they purchase.

This episode, however unglamorous compared to the more celebrated stories of scientific discovery that fill most of this book, is worth including precisely because it illustrates a theme this book returns to repeatedly: once a trace element becomes sufficiently valuable and sufficiently concentrated within an accessible physical object, human behavior around that object tends to reorganize itself around the economic incentive that value creates, for better and, in this particular case, rather conspicuously for worse. The same underlying scarcity that makes rhodium chemically indispensable inside a catalytic converter is precisely what made stealing that converter profitable enough to be worth the risk.

What remains striking, viewed from the perspective this book has been developing across its preceding chapters, is how directly the catalytic converter's success traces back to the same underlying principle explored in the story of molybdenum inside a nitrogenase enzyme and iron inside an ammonia reactor: a rare and precisely chosen element, deployed not in bulk but at a carefully engineered surface, can transform the chemistry of an entire system while contributing only a vanishing fraction of that system's total mass. The difference here is one of context rather than principle. Where nitrogenase evolved its molybdenum-centered solution across billions of years of biological trial and error, and the Haber-Bosch catalyst emerged from a few intense years of deliberate industrial experimentation, the modern catalytic converter represents something closer to a hybrid of the two: decades of applied engineering research, building on a foundation of basic chemistry, arriving at a solution precise enough to solve, with a few grams of rare metal, a pollution problem that had once seemed to require nothing less than reengineering the automobile itself.

The next chapter turns from the exhaust pipe to the kitchen drawer, tracing the story of an element discovered almost by accident by a metallurgist trying to solve an entirely unrelated military problem, one whose eventual success would put a thin, invisible, self-repairing chemical shield on everything from surgical instruments to the cutlery most people use without a second thought every single day.

Chapter 7

The Spark That Wouldn't Rust

Harry Brearley was not trying to invent stainless steel.

In May of 1912, Brearley, a metallurgist at the Brown-Firth Research Laboratories in Sheffield, England, traveled roughly one hundred thirty miles south to the Royal Small Arms Factory in Enfield, where he had been asked to investigate a stubbornly practical problem: the interior surfaces of rifle barrels were eroding too quickly, worn away by the extreme heat and pressure of repeated firing, and the British military wanted a steel alloy that could better withstand this erosion. Brearley's task had nothing to do with corrosion resistance or rust prevention, the properties that would eventually make his name synonymous with one of the most consequential materials of the twentieth century. He was chasing durability against heat and mechanical wear, an entirely different metallurgical problem.

The specific idea he pursued was to vary the chromium content of experimental steel alloys, testing whether higher chromium concentrations might improve resistance to the kind of erosion that was chewing through rifle barrels. Over the following year, Brearley produced a long series of crucible steel melts, methodically varying the chromium content across a range from roughly six to fifteen percent, none of which delivered the specific erosion-resistance results he had been asked to find. It was, by any conventional measure, a frustrating and largely unsuccessful year of research.

CAST NUMBER 1008

On August 13, 1913, working with an electric furnace rather than the crucible method he had relied on previously, Brearley poured an alloy containing approximately twelve point eight percent chromium and roughly zero point two four percent carbon. This first electric furnace cast did not immediately produce anything remarkable. It was, by most accounts, unsatisfactory. A second cast, on August 20, recorded in laboratory logs as Cast Number 1008, contained a very similar composition, twelve point eight percent chromium, zero point two four percent carbon, along with small quantities of manganese and silicon, and this time something different happened.

Brearley, testing the polished samples with the etching reagents he customarily used for metallurgical analysis, reagents based on nitric acid designed to reveal a metal's internal crystalline structure under examination, found that this particular chromium steel simply would not etch. It resisted the acid attack that would have visibly marked and corroded an ordinary steel sample almost immediately. In his later autobiography, published in 1941 under the title Knotted String, Brearley recalled how immediately this result distinguished the new alloy from every other steel he had tested. Where his other experimental melts corroded readily under acid exposure, Cast Number 1008 simply sat there, its polished surface unmarked, refusing to rust or stain under conditions that would have degraded conventional steel within minutes.

Brearley had not set out to discover a corrosion-resistant steel. He had stumbled into one, an accidental byproduct of a research program aimed at an entirely different metallurgical goal, and the discovery is worth pausing on for what it reveals about how trace element chemistry has actually advanced across history, a pattern that recurs throughout this book far more often than the tidy, deliberate narratives of scientific triumph might suggest. Molybdenum's role in biological nitrogen fixation was uncovered through systematic experimentation, and the Haber-Bosch catalyst emerged from Alwin Mittasch's exhaustive, deliberate screening of thousands of formulations. But a meaningful share of the discoveries recounted in this book, Brearley's among the clearest examples, trace back not to a hypothesis deliberately tested and confirmed but to a sharp-eyed researcher noticing that something unexpected had happened, and having the curiosity and technical background to recognize its significance rather than simply discarding a failed experiment.

THE INVISIBLE SHIELD

The scientific explanation for why chromium produces this remarkable corrosion resistance was not fully understood by Brearley himself, but subsequent research has clarified the mechanism with considerable precision, and it turns out to be an elegant example of exactly the kind of structural alteration described in this book's opening chapter. Ordinary steel, composed overwhelmingly of iron along with a small percentage of carbon, corrodes when exposed to oxygen and moisture because iron reacts readily with atmospheric oxygen to form iron oxide, the reddish brown compound commonly known as rust. This reaction, once initiated, tends to be self-perpetuating and progressive, because rust itself is a porous, flaking material that does nothing to protect the iron underneath it from further oxidation, allowing corrosion to steadily eat deeper into an exposed piece of iron or ordinary steel over time.

Chromium behaves very differently when exposed to the same oxygen and moisture. It also oxidizes, forming a compound called chromium oxide, but chromium oxide possesses a critical property that iron oxide lacks: it forms an extremely thin, dense, tightly adherent layer directly on the metal's surface, a layer so thin it is invisible to the naked eye and yet so effective as a physical barrier that it prevents any further oxygen from reaching the metal underneath. This protective layer is often described by materials scientists as a passivation layer, and its most remarkable characteristic is that it is self-healing. If the surface of a piece of stainless steel is scratched or otherwise damaged, exposing fresh metal to the air, the chromium in that freshly exposed metal immediately reacts with atmospheric oxygen to form a new protective oxide layer, effectively repairing the scratch's vulnerability to corrosion within moments, without any human intervention required.

When chromium is alloyed into steel at a sufficient concentration, typically above roughly ten to eleven percent by weight according to the metallurgical thresholds later established for defining true stainless steel, this passivation behavior extends across the entire alloy's surface, protecting not just isolated chromium atoms but the iron atoms interspersed among them as well. The underlying electrochemistry involves a comparison of what materials scientists call electrode potential between chromium and iron, with chromium's more negative electrode potential relative to iron meaning that chromium preferentially forms its protective oxide layer before iron has the opportunity to corrode on its own. In effect, a relatively modest addition of chromium, roughly one part in eight of the total alloy by weight, hijacks the corrosion chemistry of the entire material, imposing its own more favorable oxidation behavior across a structure that is still, by mass, composed predominantly of ordinary iron.

A DISCOVERY NOBODY WAS WAITING FOR

Despite the genuine significance of what Brearley had found, the initial reception from his employer was tepid at best. Firth Brown Steels, the company employing Brearley, showed little interest in the potential military or industrial applications of what he had begun calling rustless steel, having been focused primarily on armament-related metallurgy rather than corrosion resistance as a marketable property. Brearley, convinced of his discovery's broader commercial potential, particularly suggested that the new alloy might be well suited to cutlery manufacturing, an application that would address one of the more persistent minor annoyances of early twentieth century domestic life: the tendency of ordinary carbon steel knives and forks to rust and stain, requiring frequent washing, drying, and periodic scouring with abrasive stones to keep them presentable, or alternatively requiring costly nickel or silver plating that would itself eventually wear through to the corrodible base metal underneath.

This suggestion was, by most historical accounts, publicly ignored by Brearley's employers, even as the company privately sent samples of the new steel to Sheffield cutlers to gauge their professional opinion. It was not until Brearley had left Firth Brown, moving on to lead a laboratory at a different Sheffield steel works, that broader awareness of the new alloy's significance began to spread. News of the discovery eventually crossed the Atlantic, with the New York Times running a brief item in January of 1915 under the headline "A Non-Rusting Steel," describing the Sheffield invention as especially suited for table cutlery, a modest public debut for a material that would go on to reshape entire industries across the following century.

A LEGAL FIGHT OVER WHO DESERVED CREDIT

Brearley's frustration with his original employer's indifference did not end with a simple parting of ways. In the years following his departure from Firth Brown, a protracted patent dispute developed over the rights to stainless steel cutlery production, pitting Brearley and his new professional collaborators against his former employer, which had, in the interim, quietly continued its own development of chromium steel cutlery based substantially on research Brearley had originally conducted while still in its employ. The resulting legal wrangling, which dragged on for a considerable period and consumed real time and money that might otherwise have gone toward further developing and commercializing the new material, offers a sober reminder that the path from a genuine scientific discovery to its full commercial and public recognition rarely runs smoothly, even when, as in Brearley's case, the underlying discovery would eventually prove to be worth an entire global industry.

Historical scholarship examining Brearley's discovery has generally been careful to note that he was neither the first researcher to identify the corrosion-resistant properties of chromium steel alloys nor the only one working in this general area of metallurgy during the same period. At least a dozen other researchers across England, France, Germany, Poland, Sweden, and the United States had studied related chromium and nickel steel alloys in the years and decades before Brearley's 1913 discovery, with some having described corrosion-resistant properties in their own experimental results and at least a few having gone so far as to patent and even commercialize related alloys. Michael Faraday himself had experimented with chromium and iron alloy combinations nearly a century before Brearley's work. What distinguished Brearley's contribution, according to historians who have examined the full scope of stainless steel's tangled origins, was less a matter of pure scientific priority and more a matter of persistence and practical follow-through: Brearley recognized the commercial significance of what he had found, pursued its development and promotion with genuine determination even in the face of institutional indifference, and successfully connected the new material to a concrete, everyday application, cutlery, that gave it a viable path to widespread public adoption.

FROM KITCHEN DRAWER TO OPERATING ROOM TO SKYLINE

Once stainless steel began finding commercial footing, its applications expanded far beyond the cutlery market that first gave it a foothold, and the expansion illustrates something important about how a trace element innovation, once established, tends to ripple outward into applications its original discoverer never anticipated. The medical field adopted stainless steel for surgical instruments and implants, valuing its corrosion resistance in an environment where instruments must be repeatedly sterilized and where any degree of rust or corrosion would pose an unacceptable infection risk. The food service industry embraced stainless steel as what one industry account describes as the gold standard for combining strength, hygiene, and aesthetic appeal, values that made it standard equipment in commercial kitchens, food processing facilities, and eventually consumer kitchen appliances around the world. Architecture and construction found in stainless steel a building material capable of maintaining a polished, corrosion-free appearance across building facades and structural elements exposed to decades of weather, without the streaking rust stains that would have marred an equivalent structure built from unprotected carbon steel.

A landmark demonstration of this architectural potential arrived with the Chrysler Building in New York City, completed in 1930, whose distinctive crown, assembled from a specific nickel-chromium stainless steel alloy, has retained its bright, unblemished appearance across nearly a century of exposure to the city's weather with remarkably little maintenance, a striking practical demonstration, visible to millions of people who have never given the underlying chemistry a moment's thought, of exactly the kind of durable, self-renewing corrosion resistance the chromium oxide passivation layer described earlier in this chapter is capable of providing across an entire human lifetime and considerably beyond it.

Metallurgists working in the decades after Brearley's original discovery have continued refining the underlying chemistry considerably beyond his original twelve point eight percent chromium formulation, developing well over a hundred distinct grades of stainless steel by varying not just chromium content but the addition of other trace and minor alloying elements including nickel, molybdenum, and titanium, each combination tuned to optimize a different balance of corrosion resistance, strength, ductility, and cost for a specific application. Some grades incorporate nickel to improve the alloy's workability and resistance to a broader range of corrosive environments, including exposure to chloride-containing substances such as seawater, which can, under sufficiently harsh conditions, still cause a form of localized corrosion called pitting even in nominally stainless steel, a reminder that the word "stainless" describes a strong general tendency rather than an absolute, universal guarantee under every possible condition.

A METAL SHAPED BY SUBTRACTION AS MUCH AS ADDITION

There is a detail in stainless steel's story worth dwelling on because it illustrates a subtlety in trace element chemistry that the simpler framing of "add a small amount of element X and get remarkable property Y" can sometimes obscure. The corrosion resistance chromium provides is not merely a function of how much chromium is present, but of how that chromium is distributed and chemically bonded within the steel's microstructure. Carbon, present in all steel as an essential component that gives the material its hardness and strength, can, under certain heat treatment conditions, react with chromium to form chromium carbide compounds. When this happens extensively, it depletes the surrounding steel matrix of the free chromium needed to form the protective passivation layer, a phenomenon metallurgists call sensitization, which can leave a nominally chromium-rich stainless steel surprisingly vulnerable to corrosion in localized regions where chromium carbide formation has stripped away the very element meant to be protecting it.

This complication, discovered and gradually understood by metallurgists working through the decades following Brearley's initial breakthrough, led directly to the development of low-carbon stainless steel grades specifically engineered to minimize chromium carbide formation, along with specialized heat treatment protocols designed to keep the protective chromium properly distributed throughout the alloy rather than locked away in a chemically inert carbide compound where it can no longer perform its protective function. The lesson embedded in this refinement echoes a theme that runs throughout this book: a trace element's power depends not simply on its presence within a system, but on the precise chemical and physical form that presence takes, a distinction that separates elegant, functioning chemistry from chemistry that looks correct on paper but fails in practice.

Stainless steel, more than a century after Harry Brearley poured Cast Number 1008 in a Sheffield laboratory while chasing an entirely different metallurgical goal, remains one of the most visible and tangible demonstrations of trace element chemistry in ordinary daily life, present in kitchen sinks, surgical scalpels, architectural facades, and the utensils most people handle multiple times a day without a passing thought for the chemistry keeping them shiny. The next chapter turns to a trace element story of a very different character, one that unfolds not across the visible surface of a metal alloy but deep within the crystal lattice of a material so precisely purified that a single stray impurity atom, invisible even under the most powerful optical microscope, was once enough to determine whether an entire electronic device would function at all.

Chapter 8

One Part in Ten Million

Every device capable of reading these words electronically, whether a phone, a tablet, or a computer screen, contains billions of tiny electronic switches, each one made from a sliver of one of the most abundant materials on Earth, deliberately contaminated with an impurity present at a concentration so faint it would once have been considered a triumph of purification simply to detect it, let alone control it. The material is silicon, the second most abundant element in the earth's crust after oxygen, found in ordinary sand and quartz in effectively limitless quantity. The contamination that transforms it from an electrically unremarkable material into the foundation of the digital age is called doping, and understanding it requires understanding a very specific and very strange kind of trace element chemistry, one operating not in a chemical reactor or a biological cell but inside the rigid, repeating lattice of a crystal.

A LABORATORY LOOKING FOR A REPLACEMENT FOR GLASS TUBES

The story properly begins at Bell Telephone Laboratories in Murray Hill, New Jersey, in the years following the Second World War, where a research group under the direction of physicist William Shockley had been assembled to pursue solid-state alternatives to the vacuum tube, the fragile, power-hungry, heat-generating glass technology that had served as the primary means of amplifying and switching electrical signals since the early days of radio. Vacuum tubes worked, but they were bulky, they consumed considerable electricity, they generated substantial heat, and they had a frustrating tendency to burn out, much like an incandescent light bulb, after a limited operational lifespan. Bell Labs, as the research arm of the American telephone monopoly, had every commercial incentive to find something smaller, more durable, and more energy efficient to replace the vacuum tubes powering the enormous telephone switching networks the company operated.

Shockley's group, which came to include the experimental physicist Walter Brattain and the theoretical physicist John Bardeen, focused their research on semiconductors, a class of materials including silicon and germanium whose electrical conductivity falls between that of a true conductor, like copper, and a true insulator, like glass or rubber, and whose conductivity, crucially, could be manipulated and controlled in ways that a simple metal conductor's could not. Wartime research into radar technology had already advanced scientific understanding of germanium and silicon considerably, since both materials had proven useful in the crystal diode rectifiers used in early radar receiver equipment, giving Bell Labs' postwar research team a meaningful technical foundation to build on.

Shockley's own early efforts focused on what physicists call the field effect, an approach to building a semiconductor amplifier that, according to his theoretical calculations, ought to have worked but which, in Shockley's repeated laboratory attempts through 1945 and 1946, stubbornly failed to produce any usable amplification at all. Bardeen, the theoretical physicist on the team, eventually proposed an explanation for this persistent failure: electrons trapped at the surface of the semiconductor material were creating what he termed surface states, effectively shielding the interior of the material from the electric fields Shockley's design depended on, blocking the intended effect before it could ever manifest.

THE NIGHT THE TRANSISTOR WORKED

Bardeen and Brattain spent much of 1947 investigating and attempting to neutralize the effect of these troublesome surface states, working through a series of experiments that grew increasingly inventive and, in at least one documented case, genuinely strange, including an attempt by Bardeen to freeze a thin film of germanium using liquid nitrogen in an effort to lock the surface electrons in place. A significant breakthrough came in November of 1947, when Brattain, acting on a colleague's suggestion, discovered that immersing the silicon sample in an electrolyte solution could neutralize the disruptive surface state effect, a finding Shockley himself would later describe in almost electric terms, as having finally overcome what had been, for two years, an apparently impenetrable barrier.

Building on this discovery, and shifting their experimental focus specifically to germanium rather than silicon, Bardeen and Brattain constructed a device using a small strip of gold foil, carefully sliced in two to create two closely spaced but electrically separate contact points, pressed against the surface of a germanium crystal. On December 16, 1947, using this improvised point-contact arrangement, the two researchers succeeded, for the first time, in amplifying an electrical signal using a solid-state semiconductor device. A formal demonstration for Bell Labs leadership followed a week later, on December 23, 1947, a date now commemorated as the transistor's true birth, when Bardeen and Brattain spoke over a telephone circuit whose signal had been amplified not by a vacuum tube but by their quarter-inch-tall arrangement of germanium and gold foil.

The device, quickly named the transistor by Bell Labs, represented a genuine scientific triumph, but it also represented, in its original point-contact form, a device that was fragile, difficult to manufacture with consistent results, and not yet ready for the kind of reliable mass production that would eventually be needed to build it into telephones, radios, and the vast array of electronic devices that would follow. Shockley, working somewhat separately from Bardeen and Brattain in the months following the initial demonstration, developed an alternative and ultimately more practical design called the junction transistor, one that would depend, far more directly and precisely than the original point-contact device had, on a technique this chapter is centrally concerned with: the deliberate, controlled introduction of trace chemical impurities into an otherwise nearly perfect crystal of semiconductor material.

A LAB WHERE THE FUTURE OF ELECTRONICS WAS BEING TESTED WEEKLY

The environment inside Bell Labs' semiconductor group during this period is worth pausing on briefly, because it illustrates a broader pattern in how transformative trace element research actually tends to unfold in practice. Bell Labs, as the research arm of a telephone monopoly with guaranteed revenue and a long institutional time horizon, could afford to fund exploratory research without the same short term commercial pressure that might have constrained a more conventionally structured private company. Researchers were given considerable latitude to pursue basic physical questions about semiconductor behavior even when the practical payoff remained genuinely uncertain, an institutional patience that proved essential given how many false starts, dead ends, and years of incremental refinement separated Shockley's initial failed field effect experiments from the eventual commercial success of the junction transistor nearly five years later. This pattern, a well resourced institution willing to sustain years of uncertain, exploratory research before any commercially viable product emerges, recurs at several points across this book's other chapters as well, from Alwin Mittasch's multi-year, thousands strong catalyst screening program at BASF to Dorothy Hodgkin's eight year crystallographic campaign to determine vitamin B12's structure, and it suggests that sustained institutional patience, every bit as much as individual scientific brilliance, has been a genuine and recurring precondition for many of the trace element breakthroughs this book has chronicled.

WHY A "FLAW" IS SOMETIMES THE ENTIRE POINT

To understand why deliberately contaminating a crystal with trace impurities could produce something as remarkable as a functioning electronic amplifier, it helps to understand a little about the electronic structure of silicon and germanium themselves. Both elements sit in what chemists call Group 14 of the periodic table, meaning each atom has four electrons available in its outermost shell to form chemical bonds. In a pure silicon or germanium crystal, every atom forms exactly four bonds with its four nearest neighboring atoms, creating an extremely stable, tightly interlocked lattice structure in which essentially every available electron is locked into a bonding role, leaving very few electrons free to move through the material and carry an electrical current. This is precisely why pure silicon and germanium behave as poor electrical conductors under ordinary conditions, closer in practice to insulators than to true conductive metals.

Doping deliberately disrupts this otherwise perfectly balanced structure by introducing trace quantities of a different element, one whose atoms are similar enough in size to fit into the silicon or germanium lattice in place of a native atom, but different enough in their electron count to throw the local electronic balance out of equilibrium. Phosphorus, sitting one position to the right of silicon on the periodic table, has five outer electrons rather than silicon's four. When a phosphorus atom is substituted into a silicon lattice, four of its five outer electrons settle into the same kind of stable bonds that silicon atoms form with their neighbors, but the fifth electron has no bonding partner to pair with. It sits only loosely attached to its parent phosphorus atom, requiring comparatively little additional energy to break free and move through the crystal as a mobile charge carrier, dramatically increasing the material's ability to conduct electrical current. Silicon doped in this way, with an element that contributes extra, loosely bound electrons, is called n-type silicon, with the n standing for the negative charge those extra mobile electrons carry.

Boron, sitting one position to the left of silicon on the periodic table, produces essentially the opposite effect. With only three outer electrons rather than silicon's four, a boron atom substituted into the silicon lattice leaves one of its four potential bonds incomplete, creating what physicists call a hole, an absence of an electron that behaves, for most practical electrical purposes, as though it were itself a mobile positive charge carrier, since neighboring electrons can shift position to fill the gap, effectively causing the hole to migrate through the crystal in the opposite direction from how an actual electron would move. Silicon doped with boron or a similar three-electron element is called p-type silicon, with the p standing for the positive charge these mobile holes effectively carry.

The truly remarkable engineering trick, the one that makes modern transistors possible, emerges when a single piece of silicon is doped with n-type impurities in one carefully defined region and p-type impurities in an immediately adjacent region, creating what is called a p-n junction. At the boundary between these two differently doped regions, a set of electrical properties emerges that allows the resulting junction to act as a controllable gate, permitting or blocking the flow of current depending on the voltage applied, the fundamental behavior underlying every transistor, and by extension every logic circuit, every processor, and every memory chip in modern electronics.

CRYSTALS PURE ENOUGH TO CONTROL

Achieving this level of controlled doping required solving an entirely separate and, in its own way, equally demanding problem: producing silicon and germanium crystals pure enough, in their undoped state, that the deliberately introduced doping impurities could actually dominate the material's electrical behavior, rather than being swamped by uncontrolled contamination left over from the crystal's manufacturing process. This was not a trivial requirement. The concentrations of dopant atoms needed to produce useful semiconductor behavior are extraordinarily small, often in the range of one dopant atom for every ten million to one hundred million atoms of the base silicon or germanium material, meaning that any uncontrolled background impurity present at a comparable or higher concentration would have rendered the deliberate doping essentially meaningless, drowned out by chemical noise the manufacturing process itself had introduced.

A COMPARISON WORTH SITTING WITH

To appreciate just how demanding this purification requirement actually was, it helps to compare it against the purity standards of other, more familiar industrial materials. Commercial grade table salt, sold for ordinary culinary use, is typically considered acceptably pure at somewhere around ninety nine percent sodium chloride, meaning roughly one part in a hundred can consist of other minerals and trace impurities without meaningfully affecting the product's usability. Pharmaceutical grade compounds, held to a considerably higher standard because human health depends on their precise composition, are often required to reach purities of ninety nine point nine percent or better, corresponding to impurity levels measured in parts per thousand. The semiconductor grade silicon that modern electronics depends on, by contrast, is routinely purified to a standard of eleven nines, meaning ninety nine point nine nine nine nine nine nine nine nine nine percent pure silicon, corresponding to a background impurity concentration measured in parts per billion, a level of purification so far beyond the standards of nearly any other manufactured material that the comparison itself risks understating just how extraordinary an engineering achievement it represents. Achieving and reliably reproducing this level of purity at industrial scale, across the tons of silicon the global semiconductor industry now consumes annually, stands as one of materials science's genuinely underappreciated triumphs, a largely invisible prerequisite without which none of the deliberate, controlled doping this chapter has described could ever have become commercially practical in the first place.

Solving this purification challenge fell largely to the chemist Gordon Teal, working within Bell Labs' broader semiconductor research effort. Teal developed a crystal pulling technique capable of producing germanium of dramatically higher purity than had previously been achievable, growing a single, essentially defect-free crystal by slowly drawing it up from a pool of molten germanium, a process that allowed impurities to be systematically excluded as the crystal formed. Teal's technique, demonstrated in the years immediately following the original 1947 transistor breakthrough, provided the foundation of pure material that controlled, deliberate doping actually required in order to work reliably.

A further refinement arrived through the work of the metallurgist William Pfann, who developed a technique called zone refining in the early 1950s. Zone refining works by passing a narrow, molten zone slowly along the length of a solid bar of germanium or silicon, exploiting the fact that most impurities are more soluble in the molten liquid phase than in the solidified crystal phase, causing impurities to be swept along with the moving molten zone and effectively concentrated toward one end of the bar, leaving the remainder of the material progressively purer with each pass. Repeated zone refining passes could reduce background impurity concentrations to levels that, for the technology of the era, represented an extraordinary achievement in materials purification, providing semiconductor researchers with crystal stock clean enough that the deliberate, controlled doping impurities added afterward could reliably and predictably dominate the material's electronic behavior.

With sufficiently pure base material available, researchers could then move on to precisely controlled doping techniques, growing crystals from a molten pool that had specific dopant elements deliberately dissolved into it in carefully calibrated quantities, allowing the dopant concentration and, critically, the location of transitions between differently doped regions within the finished crystal to be controlled with a level of precision that would have been unimaginable just a few years earlier. This combination of ultra-pure base material and precisely controlled trace doping culminated, on April 20, 1950, in the demonstration of the first practical n-p-n junction transistor, a milestone Shockley recorded in his own laboratory notebook with characteristic understatement, simply noting that the unit had been successfully demonstrated that day.

FROM A NEW JERSEY LABORATORY TO EVERY POCKET ON EARTH

The commercial and technological consequences of this research, once fully unleashed, proved to be about as far from understated as any development in the history of technology has been. Transistors, requiring far less power, generating far less heat, and offering far greater durability and reliability than the vacuum tubes they replaced, found their way within a decade into telephones, radios, hearing aids, guided missile systems, and early pocket calculators, and the ongoing refinement of doping and crystal purification techniques over the following decades enabled the progressive miniaturization that eventually made it possible to fit not just one transistor but billions of them onto a single silicon chip smaller than a fingernail.

Modern semiconductor manufacturing has extended the basic doping principles Bardeen, Brattain, Shockley, Teal, and Pfann worked out in the years immediately following the Second World War to levels of precision that would likely have struck even those pioneering researchers as extraordinary. Contemporary chip fabrication techniques can control the placement of individual dopant atoms within a silicon crystal with nanometer-scale precision, a level of control that has allowed the semiconductor industry to continue shrinking transistor dimensions and increasing transistor counts on a single chip for decades, a trend popularly summarized by Moore's Law, itself made possible at every step by the ongoing refinement of exactly the kind of trace impurity chemistry this chapter has described.

THREE MEN, ONE PRIZE, AND A QUIET FALLING OUT

The scientific recognition that followed the transistor's invention offers its own small coda to this chapter's story, one that complicates the tidy image of a harmonious research team working in perfect collaborative unison. Bardeen, Brattain, and Shockley shared the 1956 Nobel Prize in Physics for their invention of the transistor, an honor that placed all three men's names permanently alongside one of the twentieth century's most consequential technological breakthroughs. Yet the collaborative unity implied by that shared prize concealed a considerably more strained working relationship than the public citation suggested. Shockley, whose own field effect research had initially failed to produce a working device, had grown visibly frustrated in the months following Bardeen and Brattain's successful December 1947 demonstration, reportedly working through the following weeks largely in isolation to develop his own alternative junction transistor design, a device that would not depend on Bardeen and Brattain's original point contact approach at all. Bardeen, for his part, left Bell Labs in 1951, moving to the University of Illinois, where he would go on to make a second, entirely separate set of foundational contributions to physics, co-developing the theory of superconductivity that would eventually earn him a second, unshared Nobel Prize in Physics in 1972, making him the only person in history to receive the Nobel Prize in Physics twice.

This underlying tension between Bardeen, Brattain, and Shockley is a useful reminder, echoing a point this book has made in the context of other discoveries as well, that the tidy institutional narrative surrounding a major scientific breakthrough, three colleagues, one shared prize, one harmonious achievement, frequently smooths over a messier and more human reality of competing egos, uneven credit, and genuine interpersonal friction among the very researchers whose combined work made the breakthrough possible in the first place.

There is a specific irony worth noting in the story of silicon doping that connects it back to a recurring theme of this book. Ordinary, undoped silicon, chemically pure and structurally perfect, is essentially useless for building the electronic devices that define modern life. It conducts electricity poorly and does so in an uncontrolled, unswitchable manner that offers no useful functionality at all. It is only through the deliberate introduction of what would, in almost any other context, be considered an unwanted contaminant, a trace impurity present at a concentration of perhaps one atom in ten million, that this abundant and otherwise unremarkable material becomes capable of the astonishing electronic behavior underlying every computer, phone, and digital device in existence. Perfection, in this particular corner of chemistry, turns out to be far less useful than a very carefully chosen and very precisely controlled flaw.

The next chapter turns away from the deliberate, celebrated trace element interventions this book has focused on so far, toward a story with a considerably darker arc: a case in which a trace chemical addition, embraced enthusiastically and marketed aggressively for decades, turned out to be quietly poisoning an entire population, and the decades-long struggle required to finally recognize, prove, and reverse that harm.

Chapter 9

The Element That Lied

"I'm not taking any chance whatever," the chemist declared, dipping his bare hands into a container of the chemical he had just spent three years developing. "Nor would I doing that every day."

The scene took place at a press conference in October of 1924, and the chemist was Thomas Midgley Jr., a mechanical engineer who had, three years earlier, discovered that a compound called tetraethyl lead, added to gasoline in very small quantities, could dramatically reduce the metallic rattling noise, known as engine knock, that plagued early automobile engines. Midgley's hand-washing demonstration was intended to reassure a skeptical press corps that had been asking uncomfortable questions about the safety of an additive whose manufacturing process had, by that point, already killed several workers and hospitalized dozens more. What the assembled reporters did not know, though some would learn it soon enough, was that Midgley himself had recently spent months recovering in Florida from a bout of lead poisoning contracted in his own laboratory, and that his confident public display of safety was, by any reasonable accounting, a calculated performance rather than an honest assessment of the risk.

This chapter tells a story that differs in shape from every other chapter in this book. The preceding chapters have each traced a version of the same basic arc: a trace element, deployed in a vanishingly small quantity, quietly and beneficially transforms some corner of human life, whether that corner is a nitrogen-starved field, a patient's failing blood, or a computer chip's electronic behavior. This chapter tells the story of a trace element deployed with exactly the same kind of chemical elegance and precision, and with catastrophic consequences that took nearly seventy years, and the stubborn persistence of one obstinate geochemist, to fully understand and finally undo.

A PROBLEM WORTH SOLVING, AND A SHORTCUT THAT SHOULDN'T HAVE BEEN TAKEN

Engine knock, the problem Midgley and his supervisor Charles Kettering set out to solve at General Motors' research division beginning around 1916, was a genuine engineering headache. It occurred when fuel inside an engine's cylinder ignited prematurely and unevenly rather than burning in the smooth, controlled combustion an engine was designed for, producing a distinctive metallic rattling sound, reducing engine efficiency, and causing accelerated mechanical wear over time. Solving this problem mattered enormously to an automobile industry racing to build more powerful, higher-compression engines, since knock became a more severe constraint the harder engineers pushed engine performance.

Midgley and his research team worked systematically through the periodic table searching for a compound that could be added to gasoline in small quantities to suppress this knocking behavior, and they gradually noticed a suggestive pattern: elements sitting toward the heavier end of what chemists call the carbon group, including tin and lead, showed particular promise. On the morning of December 9, 1921, chemists delivered a small quantity of tetraethyl lead to the laboratory, Midgley added it to the fuel of a single-cylinder test engine, and the knocking, which had persisted through dozens of previously tested compounds, stopped completely.

The underlying chemistry, understood with far greater clarity by researchers in the decades since, involved lead's ability to interrupt the runaway chain reactions of reactive molecules responsible for premature ignition, a mechanism sometimes described as lead acting like a chemical sponge, absorbing the volatile intermediate compounds that would otherwise trigger knock before those compounds could accumulate to a problematic concentration. It was, on its own narrow technical terms, a genuinely elegant piece of applied chemistry, precisely the kind of small trace addition producing an outsized beneficial effect that this book has celebrated in every chapter before this one. The tragedy embedded in tetraethyl lead's story is not that the underlying chemistry was somehow inferior to the chemistry described in previous chapters. It is that the element chosen to exploit that chemistry happened to be one of the most well-documented neurotoxins available to industrial chemistry, a fact that was neither unknown nor genuinely uncertain at the time the decision was made to commercialize it.

DEATHS THAT WERE EXPLAINED AWAY

General Motors, together with Standard Oil of New Jersey, the corporate forerunner of what is now ExxonMobil, and the chemical company DuPont, formed a joint venture called the Ethyl Corporation to manufacture and market the new gasoline additive, and the choice of name was itself a small, telling piece of the story: the product was marketed under the brand name Ethyl, a deliberate effort to avoid the word lead entirely in public-facing communications, despite the fact that lead was, chemically and toxicologically, the entire point of the product.

The human cost of this decision became apparent almost immediately once industrial-scale production began. At Standard Oil's Bayway refinery in New Jersey, workers manufacturing tetraethyl lead began falling seriously ill in 1924, and the outcomes were severe: thirty two workers were hospitalized and five died from acute lead poisoning, with numerous accounts describing workers experiencing violent hallucinations and psychiatric disturbance in the period before their deaths, symptoms severe enough that the affected wing of the facility earned the grim nickname "the loony gas building" among workers and journalists covering the unfolding story. A Standard Oil manager, responding to press inquiries, offered an explanation that has not aged well in the historical record, suggesting that the workers had probably suffered their symptoms because they worked too hard, a framing that shifted responsibility away from the toxic compound they were manufacturing and onto the workers' own supposed overexertion.

Midgley's own bout of lead poisoning, severe enough to require an extended recuperative absence in Florida, occurred during this same period, and it should have settled any remaining ambiguity about tetraethyl lead's acute toxicity within the small circle of people directly involved in its manufacture. It did not stop Midgley from continuing to publicly defend the product's safety, culminating in the hand-washing demonstration described at the opening of this chapter, staged specifically to counter the mounting press scrutiny and public alarm the Bayway deaths had generated.

Public health officials of the era were not uniformly silent. The United States Public Health Service issued warnings as early as 1922 regarding the dangers associated with tetraethyl lead's production and use, and several jurisdictions, including New York, Philadelphia, and the state of New Jersey, imposed temporary bans on leaded gasoline sales in 1925 in direct response to the worker deaths and mounting public concern. A 1925 public conference convened to evaluate the controversy did not seriously weigh available alternatives to tetraethyl lead, despite the fact that researchers of the era were aware that ethanol, blended into gasoline at higher concentrations, could achieve a comparable anti-knock effect without lead's toxicity, an option that industry representatives dismissed, citing the larger blending volumes ethanol would require and the correspondingly higher cost that implied. These early bans and warnings, whatever momentum they briefly generated, did not hold. Industry pressure and continued lobbying eventually reversed the local prohibitions, and leaded gasoline went on to become the standard, near-universal automotive fuel formulation across the United States and much of the industrialized world for the following half century.

THE ROAD NOT TAKEN

The alternative that regulators and industry representatives declined to pursue in 1925, ethanol blended into gasoline at a higher concentration than tetraethyl lead required, deserves a slightly closer look, because the reasons for its rejection illustrate how thoroughly commercial incentive can shape a decision that, on public health grounds alone, might otherwise have gone the other way. Ethanol offered genuine and well understood anti-knock properties, and unlike tetraethyl lead, it presented no comparable acute toxicity risk to refinery workers or, eventually, to the broader population exposed to vehicle exhaust. What ethanol lacked, from the perspective of the companies developing and marketing tetraethyl lead, was patent protection and the accompanying exclusive profit potential; ethanol was a well known, easily produced commodity that any competitor could manufacture, whereas tetraethyl lead, protected by patents held jointly by General Motors, Standard Oil, and DuPont, offered its developers a proprietary, licensable product with a far more favorable commercial return. Charles Kettering himself, according to later historical accounts, privately acknowledged ethanol's technical viability as an anti-knock additive even while publicly championing tetraethyl lead, a detail that has struck later historians examining the episode as a particularly clear illustration of how commercial incentive, rather than any genuine absence of a safer alternative, shaped one of the more consequential public health decisions of the twentieth century.

A GEOCHEMIST WHO REFUSED TO STOP ASKING QUESTIONS

The scientist most responsible for eventually forcing a genuine public reckoning with leaded gasoline's environmental and health consequences was not a toxicologist or a public health official but a geochemist named Clair Cameron Patterson, whose path to the subject arrived by way of an entirely different research problem. Patterson, working in the years following the Second World War, had been attempting to precisely date the age of the Earth using radiometric analysis of lead isotopes found in ancient meteorite samples, painstaking work that required extraordinarily clean laboratory conditions in order to detect the meteorites' genuine, ancient lead content without it being swamped by contamination from more recent, environmentally distributed lead.

What Patterson discovered, largely as an unwelcome side effect of trying to solve this entirely separate scientific problem, was that his laboratory samples, and indeed seemingly every sample he attempted to analyze, were being contaminated by ambient environmental lead at levels far higher than he had anticipated, a contamination problem so persistent and so pervasive that it forced him to develop some of the cleanest laboratory contamination-control protocols that had existed anywhere in science up to that point, techniques that would later become foundational to the entire field of trace element geochemistry.

This unwanted contamination problem led Patterson toward a much larger and more consequential question: how much lead was actually present in the natural environment, distributed globally, and how did that level compare to the amount of lead present before industrial society began releasing it into the air through leaded gasoline combustion. Patterson's subsequent research, drawing on ice cores and other environmental samples that preserved a record of atmospheric lead deposition stretching back centuries, produced landmark findings published in 1965 demonstrating a dramatic and unmistakable spike in environmental lead levels closely tracking the introduction and widespread adoption of leaded gasoline, directly challenging the lead industry's long-standing public position that background environmental lead exposure was a normal, essentially unavoidable feature of modern life rather than a specific consequence of a specific industrial decision.

Patterson's findings set him on a collision course with the lead industry that would define much of the remainder of his scientific career. Senate hearings convened in 1966 placed Patterson directly opposite Robert Kehoe, the Ethyl Corporation's own longtime toxicologist, who had spent decades serving as the industry's primary scientific authority on lead safety, a role in which Kehoe consistently maintained that no scientifically established threshold of harm had been demonstrated for the levels of lead exposure leaded gasoline produced in the general population. Patterson, testifying before the same hearings, argued forcefully that allowing an industry to conduct and control the scientific investigation into its own product's public health dangers represented what he bluntly termed a direct conflict of interest, a position that would eventually be vindicated by the accumulating weight of independent research but which, in the more immediate term, cost Patterson research funding and drew sustained industry pushback against his work and his professional standing.

WHAT SEVENTY YEARS OF EXPOSURE ACTUALLY COST

The scientific case against leaded gasoline continued to strengthen through the 1960s and 1970s, as researchers working independently of industry funding increasingly documented clear links between environmental and atmospheric lead exposure and a range of serious health effects, particularly in children, whose developing nervous systems proved considerably more vulnerable to lead's neurotoxic effects than adult physiology. Lead exposure in children came to be associated with reduced cognitive function, lower measured intelligence, and behavioral difficulties, effects that, unlike the acute poisoning and hallucinations suffered by Bayway refinery workers in 1924, often manifested subtly and cumulatively rather than in any single dramatic, easily attributable medical event, a pattern that made the case against leaded gasoline considerably harder to prove conclusively and correspondingly easier for interested parties to dispute or delay addressing.

The regulatory response, once it finally began in earnest, proceeded in gradual stages rather than a single decisive action. The United States Environmental Protection Agency began phasing down permissible lead content in gasoline through the 1970s, driven in substantial part by an entirely separate but converging concern: leaded gasoline was found to damage and disable the catalytic converters described in the previous chapter of this book, the same platinum, palladium, and rhodium-based emissions control technology the Clean Air Act had mandated for new vehicles, since lead compounds in exhaust would coat and permanently deactivate the precious metal catalyst surfaces the converters depended on. This technical incompatibility between two major regulatory priorities, reducing engine knock through leaded fuel and reducing broader vehicle emissions through catalytic converters, ultimately worked in favor of eliminating lead, since unleaded fuel was required for the catalytic converter technology to function at all.

The complete phase-out of leaded gasoline for ordinary consumer vehicles in the United States was not finalized until 1996, a full seventy five years after Midgley's original December 1921 laboratory discovery and more than three decades after Patterson's landmark 1965 research had first documented the scale of the resulting environmental contamination. The World Bank formally called for a global ban on leaded gasoline that same year, and the European Union completed its own phase-out by 2000, with the global elimination of leaded gasoline for road vehicles not achieved until Algeria, the final holdout country, ceased its use in July of 2021, exactly one hundred years after Midgley's initial discovery.

The public health consequences of this extended, seventy-year delay have been the subject of extensive subsequent research, and the general scientific consensus that has emerged is sobering. Studies examining the relationship between historical leaded gasoline exposure and childhood blood lead levels have documented reductions in average blood lead concentrations exceeding ninety percent in the United States following the phase-out, a dramatic public health improvement that stands as clear evidence of just how substantial the population-wide exposure had been while leaded gasoline remained in widespread use.

Researchers have continued working to quantify the cumulative cognitive and public health toll of the intervening decades of exposure, work that remains genuinely difficult given the challenge of establishing precise causal estimates across such a large and historically distant population, but the consistent thread running through this body of research is that the delay between Patterson's initial 1965 findings and the eventual 1996 phase-out represented decades of preventable exposure, sustained in substantial part by industry resistance to findings that threatened a profitable product line.

TWO INVENTIONS, ONE INVENTOR, ONE GRIM ACCOUNTING

Thomas Midgley Jr.'s career did not end with tetraethyl lead. He went on to develop chlorofluorocarbons, marketed under the brand name Freon, as a supposedly safer refrigerant to replace the toxic and flammable compounds previously used in refrigeration systems, an invention celebrated in its own time for solving a genuine and pressing safety problem. Decades after Midgley's death in 1944, chlorofluorocarbons would themselves be identified as the primary chemical culprit behind the depletion of the Earth's protective ozone layer, a discovery that eventually led to their own global phase-out under the 1987 Montreal Protocol. One environmental historian, reflecting on the combined legacy of Midgley's two most consequential inventions, has offered an assessment that has become something of a dark epitaph, observing that Midgley had more impact on the Earth's atmosphere than any other single organism in the planet's history, a judgment that captures, in a single sentence, the genuine scale of both achievements and the genuine scale of both their unintended consequences.

A LESSON THE NEXT GENERATION OF CHEMISTS TOOK TO HEART

The prolonged, decades-long struggle required to finally phase out leaded gasoline left a lasting mark on how toxicology and environmental regulation are practiced today, well beyond the specific case of lead itself. The methodological approach Clair Patterson pioneered, developing ultra-clean laboratory protocols specifically to distinguish an element's natural background presence from its more recent industrial contribution, became a foundational technique across the broader field of environmental geochemistry, applied since to tracking the environmental fate of numerous other industrial contaminants beyond lead. Patterson's insistence, voiced most forcefully during his contentious 1966 Senate testimony, that an industry should not be permitted to serve as the primary scientific authority evaluating the safety of its own product, has become a widely cited touchstone in subsequent debates over regulatory independence, invoked in later controversies involving tobacco, asbestos, and a range of other industrial products whose manufacturers similarly sponsored and promoted favorable safety research while independent scientists raised contrary concerns.

There is a final, quietly hopeful detail worth including at the close of this chapter, one that offers a small counterweight to its otherwise sobering account. Clair Patterson, in addition to his research into leaded gasoline, is also credited with providing the first genuinely precise calculation of the age of the Earth, dating our planet to approximately four and a half billion years old based on his analysis of lead isotopes in ancient meteorite samples, the very research project that had originally led him, entirely by accident, toward his discovery of the leaded gasoline contamination problem in the first place. It is a fitting, if unplanned, symmetry that the same meticulous attention to trace quantities of lead that allowed Patterson to date the Earth itself with unprecedented precision also equipped him with exactly the analytical tools and exactly the stubborn scientific rigor needed to eventually force a reluctant industry and a slow-moving regulatory system to confront the harm their own product had been quietly causing for decades.

This chapter closes this book's exploration of trace elements' constructive power with a necessary and deliberate note of caution, because the pattern this book has traced across its preceding chapters, a small, precisely deployed quantity of a chemical element producing an outsized effect on the systems around it, is morally neutral in itself. It is neither inherently beneficial nor inherently harmful. It is simply a description of how certain kinds of chemistry behave, an accurate account of leverage without any accompanying guarantee about which direction that leverage will ultimately be exercised. Molybdenum's leverage feeds crops. Chromium's leverage prevents rust. Lead's leverage, deployed with exactly the same kind of chemical elegance and precision that makes this book's other stories so compelling, quietly damaged the developing brains of an unknown but almost certainly substantial number of children across several decades of the twentieth century, before a persistent geochemist working on an unrelated problem happened to notice that something was deeply wrong, and before that persistence, sustained across years of industry resistance, finally forced the world to act on what the evidence had been showing all along.

The next chapter returns to more familiar territory, tracing the story of an element whose leverage, so far, has proven to be considerably more straightforwardly beneficial: lithium, the lightest solid element on the periodic table, and the decades-long scientific journey that transformed it from an obscure laboratory curiosity into the technology now quietly powering nearly every portable electronic device on Earth.

Chapter 10

The Featherweight Revolution

An element rarely gets to play the leading role in a genuine drama, but the story of lithium comes about as close as chemistry gets to a story with a clear protagonist, one whose origins, appropriately for the lightest solid element on the entire periodic table, reach all the way back to the earliest minutes of the universe itself, when lithium was among the first handful of elements forged in the aftermath of the Big Bang, long before stars existed to manufacture anything heavier.

For most of the nearly two centuries since a Swedish chemist named Johan August Arfwedson first identified lithium in a mineral sample in 1817, and Jöns Jacob Berzelius subsequently named the new element after the Greek word for stone, lithium remained a scientific curiosity with only modest practical application, valued mainly for a handful of specialized industrial and, eventually, pharmaceutical uses. That changed dramatically in the final decades of the twentieth century, when three researchers, working across separate institutions and separate decades, gradually assembled the pieces of a technology that would put a small quantity of lithium, carrying an electrical charge back and forth between two electrodes, into the pocket of a substantial fraction of the human population.

A CHEMIST AT EXXON TAKES A CHANCE ON A DIFFICULT ELEMENT

The first crucial piece of this story emerged in the 1970s, from the work of a chemist named Stanley Whittingham, employed at the time by the oil company Exxon during a period when the company, alarmed by the era's oil price shocks, was investing meaningfully in energy research extending well beyond its traditional petroleum business. Whittingham took what was, by the standards of battery research at the time, a genuinely unconventional approach, developing what would become recognized as the first functional lithium-based rechargeable battery.

The core engineering challenge Whittingham confronted was substantial. Lithium's appeal as a battery material stems directly from its extreme position on the periodic table: as the lightest of all solid elements, it offers an exceptionally favorable ratio of stored electrical energy to physical weight, and its strong tendency to readily give up its single outer electron makes it, in electrochemical terms, a highly reactive and energetic material, precisely the property that makes an effective battery material in the first place. That same reactivity, however, made lithium notoriously difficult to work with safely, since pure lithium metal reacts vigorously, sometimes dangerously, with moisture and with many of the electrolyte materials a functional battery would otherwise depend on.

Whittingham's solution centered on a compound called titanium disulfide, used as the battery's cathode material, chosen specifically because its layered crystal structure created microscopic gaps capable of accommodating lithium ions as they moved into the material during the battery's discharge cycle and then releasing those same ions back out again during recharging, a reversible process that forms the essential functional basis of any rechargeable battery. Combined with a lithium metal anode, Whittingham's design demonstrated that a genuinely rechargeable lithium battery was achievable, a foundational proof of concept that Exxon pursued toward commercialization before ultimately setting the technology aside as the oil price pressures that had originally motivated the research eased later in the decade.

A TEXAS PROFESSOR DOUBLES THE VOLTAGE

The next major advance came from John Goodenough, a physicist and materials scientist who, working first at Oxford and later at the University of Texas at Austin, took Whittingham's foundational lithium battery concept and substantially improved upon it. In 1980, Goodenough developed a new cathode material using lithium cobalt oxide, a compound whose molecular structure contains open spaces capable of housing lithium ions in a manner conceptually similar to Whittingham's titanium disulfide, but with electrochemical properties that produced a meaningfully higher battery voltage than the earlier design had achieved, a critical improvement for any battery technology intended to compete against and eventually surpass the established rechargeable battery chemistries already on the market.

Goodenough's path to this discovery reflected a genuinely unusual career trajectory, one worth pausing on briefly because it complicates any simple narrative about scientific discovery following a single, tidy, linear path. Goodenough had studied mathematics as an undergraduate at Yale, served as a meteorologist in the United States Army during the Second World War, and only afterward pursued the doctoral studies in physics that eventually led him toward materials science and battery research, a circuitous route that, rather than disqualifying him from making a foundational contribution to battery chemistry, arguably gave him an unusually broad and flexible scientific toolkit to draw on. Goodenough would go on to have one of the longest and most productive research careers in the history of materials science, continuing active laboratory work into his late nineties, a detail that becomes relevant again later in this chapter's account of how his contribution was eventually recognized.

A JAPANESE CHEMIST SOLVES THE SAFETY PROBLEM

The final essential piece of the modern lithium-ion battery came from Akira Yoshino, a chemist working at the Japanese company Asahi Kasei, who confronted directly the safety limitation that had continued to constrain lithium battery technology even after Goodenough's cathode improvements: the continued use of pure lithium metal as the battery's anode material, which remained prone to forming needle-like structures called dendrites during repeated charging cycles, structures capable of eventually piercing the thin separator material meant to keep a battery's two electrodes safely apart, creating a short circuit and, in a worst case scenario, a fire or explosion.

Yoshino's solution, developed through the first half of the 1980s and formalized in a patent application filed in 1985, replaced the unstable pure lithium metal anode with a carbon-based material, initially exploring a conductive plastic called polyacetylene before settling on other carbonaceous materials, into which lithium ions could be safely stored and released without the dendrite formation risk that pure lithium metal presented.

A CAREER THAT BEGAN WITH A CHILDHOOD BOOK ABOUT CANDLES

Yoshino has, in interviews conducted after his eventual Nobel recognition, traced his early interest in chemistry to a childhood reading of Michael Faraday's nineteenth century popular science lectures on the chemistry of a burning candle, a formative influence he has cited repeatedly as sparking the curiosity about chemical transformation that would eventually lead him, decades later, toward battery research. This detail is worth including not merely as biographical color but because it illustrates a pattern this book has touched on before in its account of John Goodenough's own winding path toward materials science: many of the researchers responsible for the trace element breakthroughs this book describes did not follow a narrowly predetermined career trajectory aimed at their eventual discovery from the outset, but instead arrived at their consequential contributions through a combination of broad early curiosity, a willingness to move between related but distinct technical fields, and a considerable amount of patient, incremental laboratory work sustained over years before any commercially significant result finally emerged.

Yoshino incorporated additional safety features into his design as well, including a heat-sensitive polyethylene-based separator film positioned between the battery's two electrodes, engineered to melt and effectively shut down the battery's internal chemical reaction if the cell began to overheat, functioning as a kind of built-in fuse that would prevent a localized malfunction from escalating into a broader failure.

Building directly on Yoshino's design, the Japanese electronics company Sony brought the first commercially successful lithium-ion battery to market in 1991, marking the technology's transition from a laboratory achievement, refined across three separate research contributions spanning nearly two decades, into a genuine consumer product ready to compete against the nickel-cadmium and nickel metal hydride battery chemistries that had previously dominated the market for portable electronic devices.

WHY THIS PARTICULAR BATTERY CHANGED EVERYTHING

The commercial success of lithium-ion battery technology, once launched, proceeded with a speed and scale that even its inventors likely did not fully anticipate at the outset, and understanding why requires appreciating just how significant an improvement lithium-ion chemistry represented over the battery technologies it displaced. Lithium-ion batteries offered substantially higher energy density, meaning considerably more stored electrical energy per unit of weight and volume, than the nickel-based battery chemistries that had previously been standard, along with meaningfully longer usable lifespans measured in charge and discharge cycles, and none of the pronounced memory effect that had plagued some earlier rechargeable battery chemistries, in which repeated partial charging cycles could gradually and permanently reduce a battery's total effective storage capacity.

A NUMBER WORTH SITTING WITH

The magnitude of this improvement is worth expressing in concrete terms rather than only in comparative generalities. Nickel cadmium batteries, the dominant rechargeable technology before lithium ion arrived, typically stored somewhere in the range of forty to sixty watt hours of energy per kilogram of battery weight. Early commercial lithium ion cells, by contrast, achieved roughly one hundred fifty to two hundred watt hours per kilogram, and subsequent decades of incremental refinement have pushed the most advanced contemporary lithium ion chemistries considerably higher still. This roughly threefold to fourfold improvement in energy density, achieved within a single generation of battery technology, is the specific numerical fact underlying the qualitative transformation this chapter has described: it is the difference between a mobile phone heavy and bulky enough to require a dedicated carrying case, as many early cellular phones were, and a phone slim enough to fit comfortably in a shirt pocket, and it is the difference between an electric vehicle with a genuinely impractical driving range and one capable of covering several hundred kilometers on a single charge, a threshold widely regarded within the automotive industry as the point at which electric vehicles became a realistic replacement for a conventional gasoline powered car for the great majority of ordinary consumer driving needs.

This combination of advantages arrived at a particularly opportune moment in the broader history of consumer electronics. The technology matured through the 1990s just as modern microelectronics was beginning to enable an entirely new generation of genuinely portable consumer devices, mobile telephones small and light enough to carry comfortably, laptop computers compact enough for regular travel, and eventually a converging wave of devices, cameras and video recorders and portable music players, that would themselves later merge into the smartphone. None of this proliferation of portable electronics would have been remotely practical using the heavier, lower-capacity battery technologies that had preceded lithium-ion chemistry. The battery, in a very real sense, had to arrive before the devices it would eventually power could be meaningfully imagined, let alone built.

The technology's significance extends well beyond consumer electronics into a domain with considerably higher stakes for the future of global energy systems. Lithium-ion batteries have become the dominant technology for electric vehicles, where their favorable energy density directly translates into practical driving range, and they play an increasingly central role in storing electrical energy generated by intermittent renewable sources like solar and wind power, holding that energy in reserve for use during periods when the sun is not shining or the wind is not blowing. This dual role, powering the portable electronics that have reshaped daily communication and information access while simultaneously enabling a meaningful reduction in fossil fuel dependence across both transportation and electrical grid storage, is precisely what the Royal Swedish Academy of Sciences cited when awarding the 2019 Nobel Prize in Chemistry jointly to Whittingham, Goodenough, and Yoshino, honoring their combined, decades-spanning contributions to a technology the prize committee described as having laid the foundation for a wireless, increasingly fossil-fuel-free society.

Goodenough, receiving the award at age ninety seven, became the oldest laureate in Nobel Prize history at the time, a distinction that generated its own share of attention alongside the underlying scientific achievement, and one that offered a fitting, almost symbolic capstone to a research career that had begun during the Second World War and continued, remarkably, well into the twenty first century.

THE COBALT QUESTION RETURNS

It is worth pausing here to note a connection this chapter shares with an earlier one in this book. Goodenough's foundational 1980 cathode material relied on lithium cobalt oxide, meaning that cobalt, the same trace element responsible for vitamin B12's remarkable biochemical potency as described several chapters earlier, also became a critical component of the battery chemistry now powering a substantial share of the world's portable electronics and a growing share of its electric vehicles. This overlap is not merely a coincidence worth noting for its own sake. It represents a genuine practical tension in the modern lithium-ion battery supply chain, since cobalt remains, notwithstanding ongoing research into cobalt-reduced and cobalt-free battery chemistries, a significant component in many of the highest-performance lithium-ion battery formulations currently in widespread commercial use, and global cobalt supply remains concentrated overwhelmingly in mining operations in the Democratic Republic of the Congo, raising the same set of ethical sourcing and labor condition concerns already discussed in this book's earlier chapter on vitamin B12 and cobalt.

Researchers working across battery chemistry laboratories worldwide continue to pursue alternative cathode formulations that reduce or entirely eliminate cobalt content, including lithium iron phosphate chemistries that have already achieved meaningful commercial adoption, particularly in electric vehicles where the tradeoff of somewhat lower energy density against reduced cost and improved supply chain security has proven commercially attractive for a growing range of applications. This ongoing search for improved formulations echoes, in its broad shape, the decades-long search for improved catalysts described in this book's earlier chapter on the Haber-Bosch process: even after a foundational trace element chemistry has been established and commercially proven, the underlying engineering and materials science work of refining that chemistry, improving its performance, reducing its cost, and addressing its supply chain vulnerabilities, continues indefinitely, driven by the same fundamental incentives that motivated the original discovery.

A CENTURY AND A HALF IN THE MAKING

There is something worth appreciating in the sheer span of time separating lithium's initial scientific discovery from its eventual transformation into a technology touching nearly every corner of modern life. Arfwedson identified the element in 1817. Whittingham's foundational rechargeable battery work did not arrive until the 1970s, a gap of roughly a century and a half during which lithium remained, for most practical purposes, a scientific curiosity of only modest commercial significance. Even after Whittingham's initial breakthrough, another two decades of refinement, spanning Goodenough's cathode improvement and Yoshino's anode redesign, were required before the technology was genuinely ready for mass-market commercial deployment.

This extended timeline offers a useful corrective to any tendency to imagine trace element breakthroughs as sudden, isolated flashes of inspiration, arriving fully formed from a single laboratory in a single moment. The lithium-ion battery, examined closely, turns out to be the product of a long, halting, multi-decade collaboration across institutions, countries, and scientific generations, each contributor building carefully on the specific limitations left unresolved by the contributors before them, until the combined result finally cleared the practical threshold required for genuine commercial success.

THE MINING QUESTION BENEATH THE BATTERY QUESTION

Lithium's transformation from a scientific curiosity into an indispensable industrial commodity has, inevitably, created its own set of extraction and supply chain challenges, ones that mirror in important respects the cobalt sourcing concerns already discussed earlier in this chapter. A substantial share of the world's lithium supply is drawn from a region spanning parts of Argentina, Bolivia, and Chile, sometimes referred to informally as the Lithium Triangle, where the metal is extracted from underground brine deposits pumped to the surface and concentrated through evaporation across vast engineered ponds. This extraction method, while less directly associated with the acute labor conditions that have drawn scrutiny to parts of the cobalt mining sector, has instead raised a different set of environmental concerns, particularly regarding its water consumption in some of the driest regions on Earth, where local and indigenous communities have raised concerns about competition between lithium extraction and the limited water resources those communities depend on for agriculture and daily life.

A separate and increasingly significant share of global lithium supply comes from hard rock mining of a lithium-bearing mineral called spodumene, concentrated particularly in Australia, which extraction method carries its own distinct set of environmental considerations related to conventional open-pit mining rather than brine evaporation. As global demand for lithium-ion batteries continues its rapid growth, driven substantially by electric vehicle adoption, researchers and mining companies alike have been investing considerably in extraction techniques designed to reduce water usage and environmental footprint, including direct lithium extraction methods that aim to recover the metal from brine more efficiently and with a smaller physical footprint than traditional evaporation ponds require. This ongoing effort to make lithium extraction more sustainable, still very much a work in progress as of this writing, represents another example of the pattern this chapter has already traced in the search for cobalt alternatives: even a broadly beneficial trace element technology tends to generate its own downstream challenges, requiring its own sustained follow-on research to fully resolve.

It is a pattern this book has already observed in the century-long refinement of stainless steel alloys following Harry Brearley's original 1913 discovery, and in the ongoing, still-unfinished search for improved ammonia synthesis catalysts nearly a century after Alwin Mittasch's original screening work at BASF. Trace element chemistry, across nearly every story this book has told, tends to reward patience and incremental refinement considerably more often than it rewards any single, isolated moment of inspiration.

The next chapter turns from an element that stores energy to one that keeps time, tracing the story of an unstable isotope of one of the most ordinary elements in all of chemistry, present in living tissue at a concentration measured in trillionths, and the ingenious insight that transformed this vanishingly rare isotope into one of the most powerful tools archaeologists and geologists have ever possessed for reconstructing the true chronology of the ancient past.

Chapter 11

The Clock Inside Everything

Somewhere in your body, right now, among the trillions upon trillions of ordinary, stable carbon atoms that make up your muscles, your bones, and every strand of your DNA, there exist a small number of carbon atoms subtly different from all the rest, atoms that are, in a very literal and continuously ongoing sense, radioactively unstable, quietly decaying at a slow but mathematically precise rate. These atoms are carbon-14, and before the middle of the twentieth century, essentially no one, including most chemists, had any practical use for them at all. Then a University of Chicago chemist named Willard Libby realized that this trace isotope, present in living tissue at a concentration low enough to be genuinely difficult to measure, could be used as a kind of clock, ticking away inside every piece of organic material on Earth, and in doing so he handed archaeology and geology one of the most transformative analytical tools either field had ever received.

AN ISOTOPE MADE BY COSMIC RAYS

To understand how carbon-14 dating works, it helps to understand where carbon-14 comes from in the first place, and the answer involves a process unfolding continuously, high above the Earth's surface, largely invisible and entirely unnoticed by the overwhelming majority of people whose bodies it is quietly supplying with trace radioactive material at this very moment.

Cosmic rays, streams of high-energy subatomic particles arriving from deep space, continuously bombard the Earth's upper atmosphere from every direction. When these cosmic ray particles collide with nitrogen atoms in the atmosphere, the most abundant atmospheric gas, as discussed at length in this book's third chapter, the collision can transform an ordinary nitrogen-14 atom into a carbon-14 atom, a nuclear transmutation that occurs at a steady, statistically predictable rate across the upper atmosphere. This freshly created carbon-14 quickly combines with atmospheric oxygen to form carbon dioxide, chemically indistinguishable in every practical respect from the far more common carbon dioxide molecules built from ordinary, stable carbon-12, and this radioactive carbon dioxide then mixes thoroughly into the broader global atmosphere, becoming available for absorption by any plant conducting photosynthesis anywhere on Earth.

Every living plant, drawing in atmospheric carbon dioxide to build its own tissue, inevitably incorporates a small, steady proportion of this carbon-14 alongside the much more abundant stable carbon-12, and every animal that eats those plants, or eats other animals that ate those plants, ends up with roughly the same proportion of carbon-14 distributed throughout its own tissue as well, since the carbon in an organism's body ultimately traces back, through however many intervening steps, to atmospheric carbon dioxide absorbed by plants. As long as an organism remains alive, continuously eating, breathing, and metabolizing, this proportion of carbon-14 to carbon-12 in its tissue remains, for all practical purposes, in a stable equilibrium with the surrounding atmosphere, since new carbon-14 is continuously being incorporated even as some of the previously incorporated carbon-14 decays away.

The moment an organism dies, however, this equilibrium breaks. No new carbon is being incorporated into its tissue from any source, but the carbon-14 already present continues to decay at its own steady, unaffected rate, gradually transforming back into nitrogen-14 through a process called beta decay.

This means that the ratio of carbon-14 to carbon-12 in a piece of once-living organic material, whether a fragment of ancient wood, a scrap of preserved textile, a bone, or a piece of charcoal from an ancient hearth, steadily and predictably declines from the moment of the organism's death onward, and because the rate of that decline is governed by well-established, extremely precise nuclear physics, measuring the remaining carbon-14 proportion in a sample allows a researcher to calculate, with meaningful confidence, how long ago that organism died.

A PHYSICIST TURNED CHEMIST, AND A DISCOVERY BORN FROM WARTIME RESEARCH

Willard Libby, born in 1908 in Grand Valley, Colorado, earned both his undergraduate and doctoral degrees in chemistry from the University of California at Berkeley, developing early specialties in the emerging field of radiochemistry, the branch of chemistry concerned with the behavior of radioactive materials, along with the associated technique of using radioactive tracers to follow chemical processes. His research career was interrupted, as so many scientific careers of his generation were, by the Second World War, during which Libby worked at Columbia University's War Research Division under the Nobel Prize-winning chemist Harold Urey, contributing to wartime research efforts before eventually joining the University of Chicago's faculty as a professor of chemistry after the war's conclusion.

Libby began the specific line of research that would lead to radiocarbon dating around 1945, drawing on earlier theoretical work by the physicist Serge Korff, who had first proposed that cosmic ray interactions with the upper atmosphere might produce carbon-14 in meaningful quantities. Libby's genuine insight, and the specific contribution that distinguished his work from the purely theoretical groundwork Korff had laid, was recognizing that this atmospheric production of carbon-14, combined with the element's participation in ordinary biological processes through the carbon cycle, created a practical, exploitable dating mechanism, one that could, in principle, be applied to any organic material regardless of its geographic origin or cultural context, offering something the field of archaeology had never previously possessed: an objective, physics-based method for estimating an artifact's age that did not depend on comparing it to other objects of presumed but uncertain relative age from the same site or culture.

Libby published his foundational theoretical paper on the subject in the June 1, 1946 issue of the journal Physical Review, and over the following several years, working with his postdoctoral associate James Arnold, he moved from theory to practical demonstration, developing the laboratory techniques necessary to actually measure the extraordinarily faint radioactivity of carbon-14 present in real organic samples. This was, in itself, a formidable technical challenge, since carbon-14 makes up only about one part in a trillion of the total carbon present in living tissue, meaning that the radioactive decay events Libby needed to detect and count occurred at a rate low enough to be easily overwhelmed by background radiation from cosmic rays, nearby radioactive materials, or even the natural radioactivity present in ordinary building materials, requiring Libby's team to develop specialized shielded detection equipment capable of reliably distinguishing the genuine, faint signal of carbon-14 decay from this pervasive background noise.

TESTING THE CLOCK AGAINST HISTORY

Libby and Arnold's crucial validation step, the one that would ultimately determine whether radiocarbon dating actually worked as a practical, reliable technique rather than remaining an elegant but unproven theoretical proposal, involved testing the method against organic samples whose ages were already known with reasonable confidence through entirely independent means, most usefully samples drawn from ancient Egyptian archaeological contexts where historical and dynastic records provided well-established chronological anchors.

Libby collaborated extensively with Robert Braidwood, an archaeologist at the University of Chicago's Oriental Institute, conducting radiocarbon tests on a range of artifacts of previously established age, including wood recovered from the casket of an Egyptian mummy and material from the deck of an ancient funeral ship. The results of these validation tests, published by Libby and Arnold in the December 23, 1949 issue of the journal Science, demonstrated a consistently close agreement between the ages calculated using carbon-14 decay measurements and the ages already established through independent archaeological and historical dating methods, providing exactly the kind of rigorous empirical confirmation needed to establish radiocarbon dating as a genuinely reliable scientific technique rather than an interesting but unproven hypothesis.

The precise numerical value Libby calculated for carbon-14's half-life, the time required for half of any given quantity of the isotope to decay, was five thousand five hundred sixty eight years, a figure that has since been refined by subsequent research to a slightly more precise value of approximately five thousand seven hundred thirty years, though the original figure, now known among specialists as the Libby half-life, remains in conventional use for certain standardized calculations within the field, a small but telling detail reflecting how deeply Libby's original foundational work became embedded in the practical conventions of an entire scientific discipline.

A REVOLUTION MEASURED IN MILLENNIA

The impact of radiocarbon dating on archaeology, once the technique became widely available to researchers beyond Libby's own laboratory, proved to be genuinely transformative, sufficiently so that historians of science routinely describe the period following its introduction as the radiocarbon revolution. Archaeologists applying the new technique to artifacts recovered from sites across both the Eastern and Western hemispheres discovered that many of the chronological assumptions underlying decades of prior archaeological interpretation required substantial revision, with radiocarbon evidence indicating that early human civilization had developed independently and roughly simultaneously across multiple, widely separated regions of the world, rather than following the more linear, single-point-of-origin diffusion model that had previously dominated much archaeological thinking, a model that had tended, not entirely coincidentally, to center Europe as the presumed source from which civilization subsequently spread outward.

A EUROPEAN TIMELINE QUIETLY REWRITTEN

The specific archaeological case that perhaps best illustrates radiocarbon dating's disruptive power involves the megalithic stone monuments of Western Europe, including Stonehenge in England and comparable stone structures across Brittany and the Iberian peninsula. Before radiocarbon dating became available, the prevailing archaeological consensus held that the technological and architectural sophistication these monuments displayed must have diffused into

Europe from the more advanced early civilizations of the Eastern Mediterranean, Egypt and Mesopotamia prominent among them, since it seemed implausible to many archaeologists of the era that comparatively isolated European societies could have independently developed the engineering knowledge such monuments required. Radiocarbon dating of organic material recovered from these European sites overturned this assumption decisively, revealing that many of the megalithic monuments actually predated the Egyptian pyramids and the great ziggurats of Mesopotamia by a significant margin, making the presumed diffusion pathway chronologically impossible and forcing a substantial rethinking of how archaeologists understood the independent development of technological and organizational sophistication across different regions of the ancient world. This particular revision, sometimes referred to by archaeologists as the second radiocarbon revolution to distinguish it from the earlier and more general adoption of the dating technique itself, illustrates just how thoroughly a single trace isotope, present in ancient organic material at a concentration of roughly one part in a trillion, could force a fundamental reassessment of assumptions that had shaped archaeological interpretation for the better part of a century.

Libby's own subsequent research using the technique he had developed produced another significant chronological revision, this one concerning the end of the last Ice Age. Analyzing wood samples recovered from beneath glacial ice deposits in North America, Libby documented that the last Ice Age had concluded approximately eleven thousand years ago, a considerably more recent date than the roughly twenty five thousand years that had previously been the prevailing scientific estimate, a substantial correction to the timeline of a major geological event, achieved entirely through the careful measurement of a trace radioactive isotope present in ancient wood at a concentration of roughly one part in a trillion.

The scientific significance of this body of work was formally recognized in 1960, when Libby received the Nobel Prize in Chemistry, honored specifically for his development of a method that had proven applicable across archaeology, geology, geophysics, and numerous other branches of science, an unusually broad citation reflecting just how far radiocarbon dating's influence had already extended beyond its original archaeological motivation within little more than a decade of its introduction.

BEYOND DIGGING UP THE PAST

Radiocarbon dating's applications have continued to expand considerably beyond its original archaeological use case in the decades since Libby's foundational work, extending into scientific domains that neither Libby nor his contemporaries likely anticipated at the technique's outset. Researchers have used radiocarbon analysis to study the functioning of the human body itself, since the same steady, predictable atmospheric production of carbon-14 that allows archaeologists to date ancient artifacts also allows researchers to track the turnover and replacement rate of cells and tissues within a living human body, providing insight into questions such as how frequently certain types of neurons are replaced over the course of a human lifetime, a question that had previously proven remarkably difficult to answer through other available research methods.

The technique has also become an important tool in climate science, where researchers use radiocarbon measurements to reconstruct historical patterns of carbon exchange between the atmosphere, the oceans, and the terrestrial biosphere, tracing how carbon has moved through the Earth's interconnected systems across recent geological history. This application acquired a particular scientific significance in the mid twentieth century, when researchers recognized that the large-scale combustion of fossil fuels, itself entirely devoid of carbon-14 because the carbon locked within coal, oil, and natural gas deposits is old enough that virtually all of its original carbon-14 content had already decayed away over the tens of thousands to millions of years since those fuels were originally formed from ancient organic matter, was measurably diluting the proportion of carbon-14 present in the modern atmosphere. This dilution effect, now generally known among researchers as the Suess effect, provided some of the earliest quantitative scientific evidence documenting the scale at which human fossil fuel combustion was altering the composition of the Earth's atmosphere, connecting Libby's original archaeological dating technique, in an application he likely never anticipated, directly to the modern scientific understanding of human-caused climate change.

FORGERY, FRAUD, AND THE LIMITS OF THE CLOCK

Radiocarbon dating's reputation for scientific authority has made it a particularly useful tool for a task its original developers likely never anticipated: exposing art forgery and archaeological fraud. Because the technique measures an absolute physical property of a sample's constituent carbon rather than relying on stylistic or contextual judgment, it has proven capable of definitively unmasking objects whose claimed age had previously rested largely on connoisseurship and circumstantial evidence. Museums and auction houses have, on numerous occasions, submitted disputed artworks and artifacts for radiocarbon testing specifically to resolve authenticity questions that art historical analysis alone could not settle conclusively, and the technique has, in a meaningful number of documented cases, revealed supposedly ancient objects to be considerably more recent fabrications, sometimes to the acute embarrassment of collectors, dealers, and institutions that had previously staked their reputations on the objects' presumed antiquity.

The technique is not, however, infallible or immune to misuse, and understanding its genuine limitations is as important as appreciating its power. Radiocarbon dating measures when an organism died, not when an object made from that organism's remains was actually crafted or used, a distinction that has occasionally misled researchers working with wood or other organic materials that may have been reused or repurposed long after their original formation. The technique also becomes progressively less precise for older samples, since after roughly ten radioactive half-lives, corresponding to a period of about fifty seven thousand years using the Libby half-life, so little carbon-14 remains in a sample that reliable measurement becomes exceedingly difficult, placing an effective outer boundary on how far back in time the method can meaningfully reach, a boundary well understood and carefully respected by the researchers who rely on the technique today.

A CLOCK RUNNING INSIDE EVERY LIVING THING

There is something genuinely striking to sit with at the close of this chapter, a fact easy to state plainly but somewhat harder to fully absorb: an isotope present in your own body at this very moment, in a concentration of roughly one part in a trillion, is continuously and silently decaying at a mathematically precise, unchanging rate, a rate entirely unaffected by anything you eat, anything you do, or anything happening in the world around you. This decay has been occurring inside every living organism, and inside your own body specifically, for as long as you have been alive, and it will continue occurring inside whatever organic remnants of you persist after death, providing, to any future researcher equipped with the right instruments, a precise physical record of how long ago that transition from living to no longer living actually took place.

This is, in its own understated way, one of the more remarkable examples in this entire book of how a trace element, present in a quantity too small for any unaided human sense to detect, can nonetheless be harnessed to answer questions of genuine and lasting significance, questions that, in the case of radiocarbon dating, reach as far back as the origins of human civilization itself and extend forward into contemporary efforts to understand and address anthropogenic climate change. Willard Libby did not discover carbon-14. Physicists working before him had already identified the isotope and theorized about its atmospheric origins. What Libby contributed, the genuine insight that transformed an obscure nuclear physics curiosity into one of the most consequential analytical tools science has ever developed, was recognizing that this vanishingly rare trace isotope, ticking away at a steady rate inside every living thing on Earth, could be read as a clock, if only someone were patient and precise enough to learn how to listen to it.

The next chapter returns from the deep timescales of archaeology and geology to a considerably more intimate and immediate subject: the human body's own daily requirement for a trace mineral whose absence, in populations across the Middle East in the mid twentieth century, produced a puzzling and previously unrecognized medical syndrome, one whose eventual diagnosis would establish an entirely new category of essential human nutrition.

Chapter 12

The Growth Metal

In 1958, a young physician named Ananda Prasad, recently arrived at a hospital in Cairo on a research assignment he had accepted somewhat reluctantly after a series of career setbacks left him with few other immediate options, was asked to examine a patient with severe iron deficiency anemia. What Prasad found, once he looked past the anemia that had prompted the referral, was considerably stranger and more puzzling than a routine case of nutritional iron deficiency. The patient, though chronologically in his twenties, had the physical stature and skeletal development of a boy of roughly ten years old. He showed no signs of sexual maturation appropriate to his actual age. His liver and spleen were both visibly enlarged. And, in a detail that would eventually prove significant to the mystery Prasad was about to unravel, the patient regularly ate clay.

This was not, as Prasad soon discovered, an isolated or unusual case. Similar patients, presenting with the same unusual combination of profound growth stunting, delayed or absent sexual development, enlarged internal organs, and clay-eating, a behavior with the clinical name geophagia, turned out to be far from rare in the region, appearing across multiple villages in patterns that suggested something systemic in the local diet or environment was responsible, rather than any individually inherited or infectious cause. Understanding what that something actually was would occupy Prasad for the following several years, and would ultimately establish an entirely new category of essential human nutrition: zinc, an element that, before Prasad's research, had never been definitively confirmed as biologically necessary for human beings at all.

A MYSTERY THAT DIDN'T FIT THE OBVIOUS DIAGNOSIS

Prasad's initial clinical instinct was to treat the iron deficiency anemia that had originally prompted the patient's referral, and this treatment did produce a genuine improvement in the patient's blood counts, confirming that iron deficiency was indeed present and was indeed responding appropriately to standard treatment. But the more striking and more medically puzzling features of the case, the profound growth stunting and the complete absence of sexual maturation despite the patient's chronological age, showed no meaningful improvement whatsoever from iron supplementation, a mismatch that told Prasad something important was being missed. Whatever was causing the growth and developmental abnormalities, it was not simply iron deficiency wearing a slightly unusual clinical presentation. Something else, some additional and as yet unidentified factor, had to be involved.

Prasad's path toward identifying that additional factor drew on research that, at first glance, seemed to belong to an entirely different scientific domain. He was aware of existing research demonstrating that zinc deficiency in laboratory rats produced a distinctive and recognizable set of symptoms, including growth failure, hair loss, thickened and abnormally textured skin, and testicular atrophy, a cluster of effects that, considered carefully, bore an intriguing resemblance to at least some of what Prasad was observing in his human patients. He was also aware of separate research documenting that phytic acid, a compound present at high concentrations in the unleavened, high-extraction wheat flour bread that formed the dietary staple of the populations he was studying, could bind tightly to dietary zinc and other minerals within the digestive tract, substantially reducing the amount of zinc the body would otherwise be able to absorb from an already zinc-limited diet.

Piecing these separate threads together into a coherent hypothesis, Prasad proposed that his patients were suffering from a previously unrecognized human nutritional deficiency of zinc, a genuinely bold hypothesis to advance at the time, given that the scientific and medical consensus of the era held that zinc, while known to be biologically important in various animal species, was distributed widely and abundantly enough in typical human diets that a genuine human zinc deficiency was considered, at best, a remote theoretical possibility rather than a real clinical concern worth actively investigating.

CONFIRMING THE HYPOTHESIS ACROSS TWO COUNTRIES

Testing this hypothesis rigorously required Prasad to move beyond a single unusual clinical case toward a broader, more systematic pattern of investigation, and his subsequent research career took him first to Iran, where he documented and studied a larger cohort of similarly affected patients, young men in their late teens and early twenties presenting with severe growth retardation, delayed sexual development, anemia despite adequate dietary iron intake, enlarged livers and spleens, and, in many cases, the same clay-eating behavior observed in his original Cairo patient. These men, drawn from poor rural villages, typically subsisted on diets consisting overwhelmingly of unleavened wheat bread with very little meat, a dietary pattern that aligned closely with the phytate-binding mechanism Prasad's hypothesis depended on, since meat represents one of the richest and most bioavailable dietary sources of zinc, and its near-total absence from these patients' diets, combined with the phytate-rich bread that dominated their caloric intake, created precisely the conditions under which a genuine zinc deficiency might plausibly develop.

Laboratory measurements confirmed the hypothesis with striking clarity. Plasma zinc levels in the affected patients averaged in the range of roughly forty two to fifty seven micrograms per deciliter, substantially below the normal reference range of above eighty micrograms per deciliter that researchers had established through comparison with unaffected individuals. Prasad's research team, using what was then a standard laboratory technique called the dithizone method for measuring zinc concentration, documented this pattern of clinically significant zinc deficiency across roughly a dozen and a half separate cases studied in detail, providing the first systematic body of clinical and biochemical evidence that a genuine, clinically meaningful human zinc deficiency syndrome actually existed.

Prasad subsequently extended this research to Egypt, collaborating with the United States Naval Medical Research Unit stationed in Cairo to study a further cohort of similarly affected young men from villages near the Egyptian capital, a study population that shared the core clinical features observed in the earlier Iranian cases while differing in certain contextual details, including a notable absence of clay-eating behavior among some of the Egyptian patients and the presence of parasitic infections, schistosomiasis and hookworm among them, that were not features of the original Iranian cases. This cross-national replication, documenting essentially the same underlying zinc deficiency syndrome across two geographically distinct populations with somewhat different secondary characteristics, provided considerably stronger evidence for the syndrome's genuine biological basis than a single population study could have offered on its own, since it demonstrated that the core zinc deficiency findings were not simply an artifact of some unrelated, geographically specific factor unique to either the Iranian or Egyptian study populations individually.

TREATMENT AS THE FINAL, CONCLUSIVE PROOF

The most persuasive evidence supporting Prasad's zinc deficiency hypothesis ultimately came not from laboratory measurements alone but from the results of direct treatment. When Prasad's research team administered zinc supplementation to affected patients, alongside a more general improvement in overall dietary quality, the results were dramatic and clinically unmistakable: patients who had previously shown no meaningful growth response despite iron supplementation and improved general nutrition began growing normally once zinc was specifically added to their treatment regimen, and delayed sexual development, previously entirely absent despite the patients' chronological ages placing them well past the expected age of puberty, began progressing normally as well.

Controlled comparison studies conducted as this research program matured further strengthened the causal case, directly comparing treatment outcomes across several different intervention groups, patients receiving no specific treatment beyond general supportive care, patients receiving dietary improvement alone without specific zinc supplementation, patients receiving iron supplementation combined with dietary improvement, and patients receiving specific zinc supplementation. Across these comparisons, zinc supplementation combined with an improved, more varied omnivorous diet consistently proved more effective at restoring normal growth than any of the alternative treatment approaches, providing exactly the kind of rigorous causal evidence, moving well beyond mere correlation or plausible biochemical mechanism, that the broader scientific and medical community would eventually require before fully accepting zinc deficiency as a genuine, clinically significant human nutritional syndrome.

A DECADE OF SKEPTICISM, FOLLOWED BY OFFICIAL RECOGNITION

Despite this accumulating body of clinical and biochemical evidence, Prasad's zinc deficiency hypothesis faced substantial and sustained skepticism from significant portions of the broader nutrition science and medical research community for nearly a decade following his initial 1958 clinical observation and his more formal published findings from 1961 through 1963. This skepticism reflected, at least in part, the deeply entrenched prior assumption, widely held among nutrition researchers of the era, that zinc's broad distribution across a wide range of common foods made a genuine human dietary deficiency implausible under all but the most extreme and unusual circumstances, an assumption Prasad's research was directly and, to many contemporaries, unexpectedly challenging.

A CAREER BUILT ON PERSISTENCE

Prasad's own account of this period, related in later interviews and retrospective writings, describes a research career that required considerable personal resilience to sustain across the years of institutional skepticism his findings encountered. Having arrived in Egypt for his original research assignment under circumstances he later described candidly as not his first professional choice, following an earlier career setback in the United States, Prasad nonetheless committed himself to following the unexpected clinical puzzle his patients presented with a rigor and persistence that would eventually prove decisive. He continued publishing his zinc deficiency findings across multiple studies and multiple patient populations throughout the 1960s even as significant portions of the nutrition science establishment remained openly doubtful, a sustained commitment to an unfashionable hypothesis that mirrors, in its broad shape, Clair Patterson's own multi-decade insistence on the significance of his lead contamination findings against considerable industry pushback, a story told in an earlier chapter of this book. Prasad would go on to build a long and productive subsequent research career centered on zinc's role in human health, eventually receiving numerous professional honors for work that, in its earliest years, had been met with considerably more institutional doubt than the belated recognition might suggest.

The scientific tide gradually turned as additional independent research groups, working with their own separate patient populations and their own separate laboratory methodologies, began publishing findings that corroborated Prasad's original observations, gradually building a critical mass of independent, converging evidence that proved considerably more difficult for skeptics to dismiss than Prasad's original findings, however carefully documented, had been on their own. This accumulating evidence culminated in 1974, when the National Research Council of the United States National Academy of Sciences formally declared zinc to be an essential trace element for human nutrition and established an official recommended dietary allowance for it, a formal act of scientific and regulatory recognition that arrived sixteen years after Prasad's original clinical observation in Cairo, a considerable delay that illustrates, in its own way, a pattern this book has already observed in other chapters: genuinely novel scientific findings, even when supported by increasingly rigorous evidence, often face a substantial period of institutional skepticism and resistance before achieving full, official recognition, particularly when those findings run directly counter to well-established prior scientific consensus.

WHAT ZINC ACTUALLY DOES, AND WHY ITS ABSENCE CAUSES SO MUCH DAMAGE

The biochemical explanation for why zinc deficiency produces such a broad and severe range of clinical effects, touching growth, sexual development, immune function, and cognitive function all at once, has become considerably clearer in the decades of research following Prasad's original discovery. Zinc, unlike some of the other trace elements discussed elsewhere in this book, does not function through a single, narrowly defined biochemical role analogous to cobalt's specific function within vitamin B12 or molybdenum's specific function within nitrogenase. Instead, zinc serves as an essential structural or catalytic component across an extraordinarily broad range of different proteins and enzymes, with current estimates suggesting that several hundred distinct human enzymes require zinc in order to function properly, along with a large family of proteins called zinc finger proteins that rely on zinc atoms to maintain the precise three-dimensional structure these proteins need in order to bind to DNA and regulate the activity of specific genes.

This unusually broad biochemical role, spanning cellular processes as fundamentally important and as functionally diverse as DNA replication, protein synthesis, cell division, and immune system function, helps explain why zinc deficiency, when it occurs, produces such a wide-ranging and severe clinical picture rather than a single, narrowly defined symptom. A deficiency in a trace element required for one specific, narrowly defined enzymatic reaction might produce a correspondingly narrow and specific clinical effect. A deficiency in an element required across hundreds of different proteins spanning multiple distinct and fundamentally important cellular processes, as zinc turns out to be, instead produces the kind of broad, multi-system clinical picture, impaired growth, delayed sexual maturation, weakened immune function, and impaired wound healing among the documented effects, that made Prasad's original patients so clinically distinctive and, in retrospect, so genuinely difficult for physicians of the era to correctly diagnose using the more conventional diagnostic frameworks then available to them.

A DEFICIENCY THAT REMAINS SURPRISINGLY COMMON

Zinc deficiency, far from being a historical curiosity confined to the specific mid-twentieth century Iranian and Egyptian populations Prasad originally studied, remains a significant global public health concern today, considerably more widespread than most people, including many people working in fields adjacent to nutrition science, generally realize. Current research estimates suggest that zinc deficiency affects a substantial share of the world's population, with commonly cited figures suggesting roughly a fifth of the global population may be affected to some meaningful degree, concentrated particularly in regions where diets remain heavily dependent on cereal grains with comparatively limited meat consumption, precisely the dietary pattern, dominated by phytate-rich grain products with limited access to zinc's more bioavailable animal-source dietary alternatives, that originally characterized the populations Prasad studied in Iran and Egypt more than six decades ago.

The public health response to this ongoing global zinc deficiency burden has drawn directly and explicitly on the biochemical understanding Prasad's original research established. Zinc supplementation has become a standard, evidence-based component of treatment protocols for childhood diarrheal disease in many parts of the world, reflecting zinc's documented and clinically significant role in supporting immune function and intestinal health, an application of the underlying research that connects directly back to the immune system effects Prasad's original patients exhibited as one part of their broader clinical presentation.

A NOSE FOR SCIENCE

Zinc's biochemical reach extends into corners of everyday human experience considerably more mundane than growth and sexual development, and one of the more curious examples involves the human sense of smell and taste. Researchers have documented that zinc plays a meaningful role in the normal functioning of taste and olfactory receptors, and clinically significant zinc deficiency has been associated with a diminished or distorted sense of taste and smell, a connection that led to zinc supplements being marketed, with somewhat mixed scientific support, as a treatment for certain cases of taste and smell disturbance, including some cases arising from viral infection. This particular application of zinc biochemistry received considerable renewed public attention during the COVID-19 pandemic, when loss of taste and smell emerged as a distinctive and widely reported symptom of infection, prompting a wave of public interest in zinc supplementation as a possible remedy, interest that outpaced, in certain respects, the more measured and still-evolving state of the underlying clinical research.

Agricultural researchers, working in a parallel research effort that echoes the earlier chapter of this book concerning molybdenum's role in nitrogen-fixing crops, have also pursued zinc biofortification strategies, developing crop varieties, wheat prominent among them, deliberately bred or engineered to accumulate higher zinc concentrations in their edible grain, an agricultural intervention specifically designed to address zinc deficiency at its dietary source in populations that continue to rely heavily on grain-based diets for the bulk of their daily caloric intake.

A DISEASE HIDDEN IN PLAIN SIGHT

There is a particular quality to Prasad's discovery that distinguishes it somewhat from several of the other stories told in this book, and it is worth drawing out explicitly as this chapter closes. The molybdenum story, the Haber-Bosch story, the stainless steel story, and the lithium battery story each concern, in their different ways, the deliberate, engineered application of a trace element to solve a problem that had already been clearly identified and understood, even if the specific solution required extended research and refinement to fully work out. Prasad's zinc research followed a meaningfully different pattern: it required first recognizing that a genuine, previously unidentified medical problem even existed at all, hiding in plain sight within a population of patients whose symptoms, considered individually, might each have been attributed to some other more familiar and better understood cause, before any question of identifying an underlying trace element deficiency, let alone developing an effective treatment for it, could even begin.

This distinction matters because it highlights a different and equally important kind of scientific contribution than the more familiar story of deliberate engineering triumph that dominates several of this book's other chapters. Sometimes the decisive scientific achievement is not cleverly engineering a novel solution to a well-understood problem, but rather noticing, with sufficient care and sufficient willingness to question prevailing assumptions, that an unusual and poorly explained pattern of human suffering, one that had likely been quietly present in affected populations for generations before Prasad's arrival in Cairo in 1958, actually reflected a specific, previously unrecognized, and ultimately quite treatable underlying cause. It took a young physician's careful clinical attention, his willingness to look past the iron deficiency that would have satisfied a more superficial diagnostic inquiry, and his persistence across nearly two decades of accumulating evidence and sustained professional skepticism, to finally establish what has become one of the more quietly consequential discoveries in the modern history of human nutrition.

The next chapter steps back from these individual element-by-element stories to trace a broader, more synthetic thread running through the entire history of metallurgy, connecting the ancient origins of alloy-making to the deliberately engineered materials science of the present day, before the book's final chapter turns forward, toward the trace elements that may prove to define the next chapter of human technological development still to come.

Chapter 13

The Alloy of Empires

Long before anyone possessed the chemical vocabulary to describe what they were doing, a metalworker somewhere in the ancient Near East, likely in the region now occupied by Iran, discovered that copper, a soft, workable, but ultimately unimpressive metal for tool and weapon making on its own, could be transformed into something considerably harder and more useful by melting it together with a small quantity of a second metal. That discovery, made independently and repeatedly across multiple regions of the ancient world sometime in the fifth millennium before the common era, marks, as much as any single moment can, the beginning of deliberate trace element metallurgy, a human tradition considerably older than the periodic table itself, older than the concept of an element in the modern chemical sense, and yet already operating, however unconsciously, on precisely the principle this book has traced across its preceding chapters: a small addition, transforming the properties of a much larger bulk material out of all proportion to its own quantity.

This chapter steps back from the individual element-by-element stories that have filled this book so far to trace a broader thread, connecting the earliest deliberate alloying experiments of the ancient world to the far more precisely engineered materials science of the present day, and in doing so, drawing out a pattern that recurs across roughly seven thousand years of human metalworking history.

AN ACCIDENT THAT BECAME A DELIBERATE CHOICE

The very earliest bronze objects recovered by archaeologists, some dating back as far as roughly 4500

BCE at sites including Pločnik in modern-day Serbia, appear to have arisen not from any deliberate alloying decision but from an accident of geology. Copper ore deposits in certain regions occur naturally alongside arsenic-bearing minerals, and early metalworkers smelting this ore, long before they had any conceptual framework for distinguishing one metallic element from another, ended up producing copper contaminated with small quantities of arsenic without having intended to do so at all. Archaeological analysis has found that the earliest such objects typically contain arsenic or tin at concentrations below roughly two percent, a level consistent with unintentional contamination carried over from the specific ore body being smelted rather than any deliberate additive decision on the part of the metalworker.

What makes this story relevant to this book's broader argument is what happened next: ancient metalworkers, presumably through repeated observation over generations, noticed that copper smelted from certain ore sources consistently produced a harder, more easily cast final product than copper smelted from other sources, and they began deliberately seeking out and combining arsenic-rich materials with their copper, whether by co-smelting copper ore alongside arsenic-bearing minerals such as realgar or orpiment, or by other techniques archaeologists have identified through careful chemical analysis of surviving artifacts. This transition from accidental to deliberate arsenic alloying, identifiable in the archaeological record once arsenic concentrations in analyzed artifacts rise consistently above roughly one to two percent by weight, a threshold researchers use specifically to distinguish intentional alloying from incidental contamination, represents an early and genuinely remarkable achievement in empirical materials science, achieved entirely without any theoretical understanding of atomic structure, chemical bonding, or any of the concepts a modern reader would consider prerequisite to understanding why the addition worked.

Arsenical bronze, once developed, spread widely across the ancient world, appearing in archaeological contexts from the Iranian plateau where it first developed, across ancient Mesopotamia, and eventually as far afield as bronze daggers recovered from the Caucasus region and, in an entirely separate and independently developed tradition, copper bells cast in ancient Mexico. Of the tens of thousands of prehistoric copper alloy artifacts catalogued in major archaeometallurgical databases, researchers estimate that roughly a third qualify as arsenical bronze by the standard compositional threshold, with a meaningful fraction containing arsenic concentrations well above two percent, evidence of just how widespread and how deliberate this particular trace element alloying practice had become across the ancient world.

A POISONOUS SOLUTION, AND THE SEARCH FOR SOMETHING SAFER

Arsenical bronze, for all its genuine metallurgical advantages over unalloyed copper, carried a serious and ultimately unsustainable cost, one that the ancient metalworkers producing it were, in all likelihood, only dimly aware of even as they suffered its consequences. Arsenic is toxic, and the smelting and working of arsenic-bearing copper ores would have exposed ancient metalworkers to arsenic fumes and dust on a regular, sustained, occupational basis, producing symptoms of chronic low-level arsenic poisoning that likely accumulated gradually across years of repeated craft work, a slow, insidious occupational hazard considerably harder to trace to its true cause than an acute, obviously connected injury would have been.

The historical record suggests that ancient metalworking communities eventually recognized, even without any formal toxicological understanding, that something about the arsenic bronze production process was genuinely harmful to the smiths who practiced it, and this recognition appears to have contributed to a gradual, multi-century shift away from arsenic and toward an alternative alloying element that could achieve broadly comparable metallurgical improvements without the same occupational health cost: tin. Tin bronze, made by alloying copper with tin rather than arsenic, offered comparable improvements in hardness, casting behavior, and corrosion resistance relative to unalloyed copper, without exposing metalworkers to the same chronic toxic hazard that arsenic bronze production carried.

The transition to tin bronze, however, came with its own considerable practical cost, one rooted not in health and safety but in geography and logistics. Tin ore, unlike the arsenic-contaminated copper ore that had made arsenical bronze production relatively straightforward wherever suitable copper deposits existed, is considerably rarer and more geographically concentrated, requiring dedicated mining and long-distance trade to reach the population centers of the ancient Near East and Mediterranean world where bronze production was concentrated. Much of the tin used across ancient Europe, remarkably, appears to have originated from mines in Cornwall, in the far southwestern corner of modern England, meaning that a functioning long-distance trade network, spanning a substantial fraction of the length of the European continent, had to develop and persist across roughly two millennia in order to supply Mediterranean and Near Eastern bronze producers with the tin their craft depended on. By around 2500 BCE, tin had become the overwhelmingly preferred bronze-alloying element across most of the ancient world that had access to it, and the elaborate trade networks required to move tin from a small number of geographically concentrated source regions to the much broader geography of bronze-consuming civilizations represent one of the earliest clear examples in human history of a critical materials supply chain, a concept this chapter's closing section, and the book's final chapter, will return to in a distinctly modern context.

THE COLOR OF TRACE ELEMENTS

Ancient metallurgists were not the only craftspeople working, centuries and millennia before the discovery of the periodic table, with trace element chemistry they could not have explained in modern terms. Glassmakers across the ancient and medieval world discovered, through generations of empirical trial and error broadly parallel to the ancient bronze workers' discovery of arsenic and tin alloying, that tiny quantities of various metal oxides, added to an otherwise clear molten glass mixture, could produce a striking and reliable range of colors: small amounts of cobalt oxide producing a deep blue, copper oxide producing green or, under different furnace conditions, a vivid red, manganese oxide serving alternately as a colorant in its own right or, in different concentrations, as a decolorizing agent capable of counteracting the pale green tint that iron impurities would otherwise impart to ordinary glass.

ROMAN GLASSMAKERS AND A TRICK REDISCOVERED TWICE

Roman glassmakers, working across the first few centuries of the common era, developed a particularly sophisticated version of this trace element manganese trick, using manganese dioxide, sourced from a mineral the Romans called sacon or, more evocatively, glassmakers' soap, in small enough quantities to counteract the greenish tint iron impurities would otherwise impart to their raw glass, producing the notably clear, colorless glass that characterizes some of the finest surviving Roman glassware. This particular application of trace element chemistry appears to have been lost, along with a considerable amount of other Roman technical knowledge, following the collapse of the Western Roman Empire, and clear glass production in Europe would not fully recover a comparable level of sophistication for well over a thousand years, until Venetian glassmakers on the island of Murano rediscovered comparable decolorizing techniques during the medieval period, developing their own closely guarded manganese based formulations that would make Murano glass one of the most prized and jealously protected trade secrets of the Venetian Republic, protected by law under penalty of death for glassmakers who attempted to leave the republic and share their techniques with foreign competitors.

This tradition of empirical, trial-and-error trace element craft knowledge, refined and passed down across countless generations of metalworkers and glassmakers long before anyone possessed a theoretical chemical framework to explain why any of it worked, represents an important and often underappreciated thread in the broader history this book has been tracing. The scientists profiled in this book's other chapters, Alwin Mittasch screening thousands of catalyst formulations at BASF, Harry Brearley noticing an unexpectedly rust-resistant sample in his Sheffield laboratory, Gordon Teal developing crystal-pulling techniques at Bell Labs, worked within a modern scientific tradition of deliberate, hypothesis-driven experimentation, informed by an established theoretical understanding of atomic structure and chemical bonding that simply did not exist for the ancient metalworkers and glassmakers who preceded them by thousands of years. And yet the underlying empirical pattern each group was working with, however differently they understood or articulated it, remains recognizably the same: small, carefully chosen additions producing large, disproportionate, and repeatable changes in the properties of a much larger bulk material.

FROM CRAFT KNOWLEDGE TO PREDICTIVE SCIENCE

What genuinely distinguishes the modern era of trace element metallurgy and materials science from its ancient predecessor is not the underlying chemical principle, which has remained essentially constant across seven thousand years of human metalworking, but rather the transition from purely empirical, trial-and-error craft knowledge toward a predictive, theory-guided science capable of anticipating which trace additions might produce useful effects before they are ever physically tested in a laboratory or furnace.

This transition did not happen instantly or completely, even within the modern scientific era this book has otherwise focused on. Alwin Mittasch's screening of thousands of potential ammonia synthesis catalysts, described in this book's third chapter, remained, for all its methodical scientific rigor, fundamentally an empirical search, testing formulation after formulation against real experimental results rather than deriving the optimal catalyst composition from first-principles theoretical calculation alone. Harry Brearley's discovery of stainless steel's corrosion resistance, described in the seventh chapter, was substantially more serendipitous still, an unanticipated observation made in the course of research aimed at an entirely different metallurgical problem. Even Gordon Teal's crystal purification work for early transistor development, while grounded in a considerably more sophisticated theoretical understanding of semiconductor physics than either Mittasch or Brearley possessed, still required extensive empirical refinement and iteration before it reliably produced the ultra-pure crystals the emerging semiconductor industry needed.

Contemporary materials science has moved considerably further along this spectrum toward genuine predictive capability, employing computational modeling techniques that can simulate, at the level of individual atoms and their electronic interactions, how a proposed trace element addition might alter a material's structural, electronic, or catalytic properties before any physical sample is actually synthesized and tested in a laboratory. This computational approach has meaningfully accelerated the pace of materials discovery across multiple domains directly relevant to earlier chapters of this book, including the ongoing search for improved ammonia synthesis catalysts capable of operating at lower temperature and pressure than the century-old iron-based Haber-Bosch catalyst, and the search for battery cathode materials capable of reducing or eliminating dependence on cobalt while preserving the high energy density that has made lithium-ion battery technology so transformative.

Even with these considerably more sophisticated modern tools, however, genuine empirical serendipity, the same basic pattern of careful observation and unexpected discovery that guided Brearley toward stainless steel in 1913, continues to play a meaningful role in materials science research, a humbling reminder that even the most theoretically sophisticated predictive models remain, in practice, approximations of an underlying physical and chemical reality complex enough to continue producing genuine surprises for researchers equipped with vastly more powerful theoretical and computational tools than any of this book's earlier subjects ever had access to.

A CONTINUOUS THREAD, SEVEN THOUSAND YEARS LONG

There is a genuine and, on reflection, rather moving continuity connecting the anonymous metalworker who first deliberately added arsenic-bearing ore to a copper smelt in ancient Iran roughly seven thousand years ago to the twenty first century materials scientist running a computational simulation to predict the electronic properties of a proposed new battery cathode material. Both are engaged in fundamentally the same intellectual project: identifying which small, carefully chosen additions to a larger bulk material will reliably produce a valuable and repeatable change in that material's behavior.

DAMASCUS STEEL AND A RECIPE LOST TO TIME

No account of ancient trace element metallurgy would be complete without mentioning one of its most famously mysterious chapters: the legendary Damascus steel of the medieval Islamic world, prized for centuries across Europe and the Middle East for a combination of exceptional sharpness, flexibility, and a distinctive rippling pattern visible on the surface of a finished blade. Damascus steel was forged, historians and metallurgists have established, from a particular imported raw material called wootz steel, produced originally in South India and Sri Lanka through a distinctive smelting process, and blacksmiths in Damascus and elsewhere then forged this imported wootz into blades whose reputation for quality spread across multiple continents through trade and, not infrequently, through the battlefield encounters of the Crusades.

The exact technique for producing authentic Damascus steel was, for reasons that remain only partly understood, lost sometime around the eighteenth century, as the supply of the specific wootz steel it depended on dwindled and the specialized smithing knowledge required to work it was not fully passed down to subsequent generations. Modern metallurgical analysis of surviving original Damascus steel blades has revealed a genuinely striking detail: some samples contain trace quantities of carbon nanotubes and cementite nanowires, structures whose existence would not be theoretically understood or even conceptually imagined by materials science for another two centuries, suggesting that the original wootz smelting process, quite possibly through trace impurities present in the specific ores used, particularly vanadium and other minor elements, inadvertently produced nanoscale structural features that modern researchers only fully appreciated once electron microscopy became available to examine the old blades at a level of magnification medieval smiths could never have accessed themselves.

This is, in its own way, one of the more remarkable illustrations in this entire book of a theme that has recurred throughout: skilled human craftspeople, working entirely through empirical observation and generations of accumulated practical experience, can sometimes stumble upon and reliably reproduce genuinely sophisticated trace element chemistry, chemistry whose full scientific explanation would not become available until long after the original craft knowledge itself had already been lost.

This continuity matters for more than merely historical or philosophical interest. It suggests that the broader pattern this book has traced across its many individual chapters, small quantities of carefully chosen elements exerting large and disproportionate effects on the materials and systems around them, is not a peculiarity of the particular scientific era this book has mostly focused on, the roughly one hundred fifty years spanning Beijerinck's discovery of biological nitrogen fixation through the most recent advances in lithium battery chemistry. It is instead a much older and more fundamental feature of how the physical world actually works, one that human beings have been quietly exploiting, with steadily improving sophistication and steadily improving theoretical understanding, for the entire span of recorded civilization, and very likely for a considerable stretch of unrecorded prehistory before that.

The final chapter of this book turns forward rather than backward, examining a family of trace elements whose importance to contemporary technology has grown dramatically in just the past few decades, elements whose supply chains and geopolitics increasingly echo, in a strikingly direct way, the ancient bronze-age tin trade this chapter has just described, and whose future availability may prove to be one of the defining material constraints shaping the technological trajectory of the twenty first century.

Chapter 14

The Elements of Tomorrow

Somewhere on a factory floor, most likely in China, which currently accounts for roughly ninety four percent of global permanent magnet production despite mining only around seventy percent of the world's raw rare earth ore, a technician is overseeing the manufacture of a magnet that will eventually spin inside the motor of an electric vehicle or the generator of a wind turbine. That magnet's exceptional strength, strength that allows it to be considerably smaller and lighter than older magnet technologies could achieve for the same power output, depends on a family of elements most people have never heard of and could not confidently place on a map of the periodic table: neodymium and praseodymium primarily, with smaller additions of dysprosium and terbium blended in to help the finished magnet retain its strength at the high operating temperatures a real-world electric motor generates.

This chapter closes the book by looking forward rather than backward, examining a cluster of trace element stories still very much in progress, whose outcomes remain genuinely uncertain in a way that the largely settled historical narratives of the preceding chapters are not. It is, in a sense, the chapter this book's author had the least certainty about while writing it, precisely because it concerns technologies and supply chains whose future trajectory has not yet been written.

SEVENTEEN ELEMENTS, ONE MISLEADING NAME

The rare earth elements, a group of seventeen chemically similar metals occupying a distinctive block near the bottom of most periodic table diagrams, carry a name that has confused generations of students and journalists alike, because most of them are not, in fact, particularly rare in an absolute geological sense. Cerium, the most abundant of the rare earths, is roughly as common in the Earth's crust as ordinary copper. What genuinely constrains rare earth supply is not underlying geological scarcity but a considerably more specific and more solvable problem: these elements rarely occur in concentrated, easily mined deposits, and separating the various rare earth elements from one another, since they share such similar chemical properties, requires an extensive, technically demanding, and often environmentally challenging refining process, one that has become concentrated in a small number of countries, China foremost among them, that have made the sustained industrial investment necessary to develop world-class rare earth separation and refining capacity.

A NAME BORN OF NINETEENTH CENTURY CONFUSION

The word "rare" in rare earth elements is, appropriately enough for a book devoted to the gap between an element's true significance and its apparent quantity, something of a historical accident rather than an accurate scientific description. The name dates to the late eighteenth and nineteenth centuries, when the first rare earth minerals were identified in a small number of unusual mineral deposits, most famously a quarry near the Swedish village of Ytterby, whose name would eventually lend itself to no fewer than four separate rare earth elements: yttrium, ytterbium, terbium, and erbium, an unusually large naming tribute to a single small mining location that reflects just how much of the foundational nineteenth century rare earth chemistry was worked out using mineral samples drawn from that one Swedish site. At the time these elements were first isolated, the mineral ores containing them did indeed seem genuinely rare and unusual, since chemists of the era had not yet identified the more widespread, if more thinly distributed, deposits that would later prove rare earth elements to be considerably more common in the Earth's crust overall than their name suggests. The name stuck long after the underlying premise it was based on had been superseded by improved geological understanding, a small but telling example of how scientific nomenclature, once established, tends to persist well past the point where it remains fully accurate.

This concentration of refining capacity, considerably more than any raw geological scarcity, is what has made rare earth elements a subject of growing strategic and geopolitical concern over the past two decades. When China imposed rare earth export restrictions in 2011, the price of neodymium climbed by more than sevenfold within a relatively short period, and the price of dysprosium rose by a figure researchers have placed at roughly two thousand percent, a demonstration, on a genuinely global economic scale, of exactly the kind of outsized leverage this book has traced across smaller and more localized examples in its preceding chapters. A small number of elements, present in the finished products that depend on them at concentrations measured in grams rather than kilograms, turned out to be capable of disrupting global supply chains for electric vehicles, wind turbines, and a wide range of other advanced technologies, precisely because so much of the world's capacity to refine and process these elements had come to be concentrated in one country's hands.

THE MAGNET THAT MAKES ELECTRIC MOTORS POSSIBLE

Understanding why rare earth elements have become so central to contemporary clean energy technology requires understanding a specific and genuinely elegant piece of applied physics. Electric motors, whether powering a vehicle or generating electricity inside a wind turbine, rely on the interaction between magnetic fields and electrical currents to convert between electrical and mechanical energy, and the strength of the permanent magnets used inside these motors directly determines how much power a motor of a given physical size can generate, or conversely, how physically small and light a motor can be built while still delivering a required amount of power.

Neodymium iron boron magnets, commonly abbreviated NdFeB, represent the strongest class of permanent magnet currently in wide commercial production, and their exceptional magnetic strength depends specifically on the neodymium and praseodymium content within the alloy, elements whose particular atomic and electronic structure gives them an unusually strong capacity to sustain a powerful, stable magnetic field. Dysprosium and terbium, added to the base NdFeB formulation in smaller quantities, address a specific practical limitation of the base alloy: NdFeB magnets, on their own, tend to lose magnetic strength as they heat up, a genuine problem for a magnet expected to operate continuously inside a motor that generates substantial heat during normal operation. Small additions of dysprosium and terbium meaningfully improve the magnet's resistance to this heat-induced demagnetization, allowing NdFeB magnets to maintain reliable performance inside the compact, high-power motors that modern electric vehicle and wind turbine design increasingly depend on.

The scale of projected future demand for these elements, driven by the ongoing global transition toward electric vehicles and renewable energy generation, is genuinely striking. The International

Energy Agency has projected that demand for the core magnet rare earth elements, neodymium, praseodymium, dysprosium, and terbium, has already doubled since 2015 and is expected to expand by roughly a further third by 2030 under current policy trajectories, while some longer-range industry projections suggest overall rare earth demand growth in the range of four hundred to six hundred percent over the coming decades. Separate research modeling the specific requirements of a full-scale global transition to renewable energy technology has estimated that meeting projected dysprosium demand alone for a single generation of green energy infrastructure would require a supply roughly two hundred times greater than actual global dysprosium mining production as of 2018, a stark illustration of just how significant the gap between current supply capacity and projected future demand has already become for at least some of the elements in this critical family.

A GEOPOLITICAL ECHO OF THE BRONZE AGE TIN TRADE

There is a genuinely striking parallel between the contemporary rare earth supply situation and the ancient bronze age tin trade described in this book's previous chapter, one worth drawing out explicitly because it illustrates how consistently this particular pattern, a critical trace material concentrated in a small number of distant source regions, driving the emergence of long-distance trade relationships and, inevitably, strategic vulnerability, has recurred across human history. Ancient Mediterranean and Near Eastern civilizations depended on tin mined in distant Cornwall, requiring an elaborate, sustained trade network spanning much of the length of the European continent to keep their bronze production supplied. Contemporary electric vehicle and wind turbine manufacturers depend on rare earth elements refined overwhelmingly in China, requiring their own elaborate, sustained, and increasingly strategically fraught global supply chain to keep their production supplied.

The response to this contemporary vulnerability has followed a broadly familiar historical pattern as well, with governments and companies outside China moving to reduce their dependence through a combination of new mining and refining investment and dedicated research into alternative materials and technologies. The United States, the European Union, the United Kingdom, Japan, and, as of relatively recently, New Zealand have all formally designated rare earth elements as critical materials warranting dedicated strategic attention and policy support, joining a growing list of countries and trading blocs treating rare earth supply security as a genuine matter of national economic and strategic policy, not merely an ordinary commercial supply chain concern to be left entirely to private market forces.

RESEARCH RACING TO REDUCE THE NEED

Parallel to these supply-side policy efforts, materials scientists have been pursuing research aimed at reducing the clean energy sector's underlying dependence on the most supply-constrained rare earth elements, an effort that echoes, in its broad shape, the decades-long search for improved and less resource-intensive catalysts described in this book's earlier chapters on ammonia synthesis and lithium battery chemistry. Researchers at the Critical Materials Institute, a federally supported research consortium based at Ames Laboratory, have been investigating magnet formulations based on cerium and cobalt, since cerium, unlike the more supply-constrained neodymium, praseodymium, dysprosium, and terbium, is one of the more genuinely abundant rare earth elements, occurring as a byproduct of processing the more sought-after magnet rare earths in quantities that have historically exceeded readily available commercial demand.

This research direction reflects a subtlety in the rare earth supply picture worth understanding, one that adds a further layer of complexity beyond the headline story of overall scarcity. Because different rare earth elements naturally co-occur within the same ore deposits and must be extracted together during initial processing, the market for the specific elements in highest demand for magnet applications cannot simply expand in isolation. Increased mining and refining activity aimed at meeting growing demand for neodymium and praseodymium inevitably produces, as an essentially unavoidable byproduct, a corresponding increase in the supply of other, less immediately sought-after rare earth elements extracted from the same ore, cerium prominent among them, creating a genuine economic incentive to develop new commercial applications, including the cerium-cobalt magnet research just described, capable of absorbing this byproduct supply rather than allowing it to accumulate as an unwanted surplus.

Beyond alternative magnet chemistry, researchers and manufacturers have also been investing considerably in rare earth recycling, developing techniques to recover neodymium, dysprosium, and other magnet rare earths from end-of-life products, ranging from retired wind turbines and electric vehicle motors to more mundane sources such as discarded hard disk drives, which have historically relied on smaller NdFeB magnets of their own. Industry analysts have projected that secondary, recycled sources could plausibly supply well over a billion dollars worth of critical rare earth elements annually by the middle of the twenty first century, a meaningful, if still comparatively modest, supplementary source of supply that could help ease, without fully resolving, the underlying tension between rapidly growing demand and geographically concentrated primary supply.

A SECOND FRONT: MOTORS THAT DO NOT NEED MAGNETS AT ALL

A separate and in some respects more radical research direction has pursued not a substitute rare earth formulation but an entirely different motor architecture that avoids the need for permanent magnets altogether. Switched reluctance motors and certain induction motor designs, both of which generate their magnetic fields using ordinary electrical windings rather than permanent magnets, have attracted renewed engineering attention specifically because of the rare earth supply concerns described throughout this chapter, despite these motor designs generally requiring more sophisticated control electronics and, in many current implementations, delivering somewhat lower efficiency than a comparably sized permanent magnet motor. Some electric vehicle manufacturers have begun incorporating rare earth free or rare earth reduced motor designs into at least a portion of their vehicle lineups, a hedging strategy that trades away some of the performance advantage rare earth magnets provide in exchange for reduced exposure to the supply chain and geopolitical vulnerabilities this chapter has described.

This diversification of technical approaches, some researchers chasing better rare earth alternatives, others chasing motor designs that avoid the need for rare earths in the first place, mirrors a pattern this book has already observed in the parallel search for cobalt reduced battery chemistries described in the tenth chapter. Facing a genuine and well-documented supply constraint on a critical trace element, engineers and researchers rarely converge on a single solution. Instead, they tend to pursue multiple, only partially overlapping strategies simultaneously, a form of technological hedging that, whatever its inefficiencies, has repeatedly proven more resilient over the long run than betting an entire industry's future on any single anticipated breakthrough.

BACK TO THE BEGINNING: A CLEANER WAY TO FIX NITROGEN

This book's third chapter told the story of the Haber-Bosch process, the century-old iron-catalyzed industrial reaction responsible for feeding a substantial share of the modern world's population, and closed by noting the ongoing, still-unresolved search for an improved catalyst capable of achieving comparable ammonia yields under less extreme, less energy-intensive conditions than the traditional process requires. It is fitting, in a book that has traced so many parallel and interconnected trace element stories, to return briefly to that unfinished search as this final chapter draws toward its own close.

The environmental stakes of this ongoing research have grown considerably clearer in recent years. The Haber-Bosch process, current estimates suggest, continues to consume roughly two percent of global energy production and is responsible for emitting more than five hundred million metric tons of carbon dioxide annually, a substantial contribution to global greenhouse gas emissions from a process whose fundamental chemistry, iron catalyst, high temperature, high pressure, has remained essentially unchanged since Carl Bosch and Alwin Mittasch's original industrial breakthrough in the years before the First World War. Researchers pursuing what has become known as green ammonia synthesis have been exploring several distinct alternative approaches, most centrally electrochemical nitrogen reduction, a technique that uses electrical current, ideally sourced from renewable generation such as solar or wind power, to directly drive the nitrogen-to-ammonia reaction under far milder temperature and pressure conditions than the traditional Haber-Bosch process requires, potentially eliminating the process's substantial fossil fuel energy input entirely.

This research remains, as of this writing, considerably short of commercial viability at the scale required to meaningfully displace conventional Haber-Bosch production, constrained by persistent technical challenges including low overall efficiency and a stubborn tendency for the desired nitrogen reduction reaction to be outcompeted by an unwanted side reaction that simply produces hydrogen gas rather than the intended ammonia. Researchers are pursuing a range of strategies to address these limitations, including the deliberate design of catalysts using earth-abundant materials structured at the single-atom level, and complementary approaches involving photocatalysis, which uses sunlight directly to help drive the nitrogen reduction reaction, echoing, in its broad conceptual approach, the way plants themselves use captured sunlight to power their own biochemistry. Whether any of these approaches will eventually mature into a genuine large-scale replacement for the century-old iron-catalyzed Haber-Bosch process remains a genuinely open scientific and engineering question, one that will likely be answered, one way or another, within the working lifetime of many people reading this book.

WHAT THIS BOOK HAS BEEN TRYING TO SHOW

It is worth pausing, at the close of this final chapter, to draw together the thread that has run continuously through every story this book has told, from the molybdenum atom sitting at the heart of an ancient bacterial enzyme to the neodymium atoms inside a twenty-first century wind turbine magnet, from the iodine quietly protecting the developing brains of Midwestern schoolchildren in the 1920s to the carbon-14 isotope silently ticking away inside every living organism on Earth right now, including, in a small and continuous way, inside your own body as you finish reading this sentence.

The thread is this: the material world does not distribute its leverage evenly. It concentrates enormous influence into vanishingly small packages, a few parts per million here, a single atom at the center of an elaborate protein scaffold there, and the difference between a civilization that understands and can deploy that concentrated leverage and one that cannot is, again and again, the difference between abundance and scarcity, between health and disease, between technological stagnation and genuine transformation. This has been true since an anonymous metalworker in ancient Iran first noticed that a particular ore produced unusually hard copper, and it remains true today, in research laboratories working to reduce the world's dependence on cobalt-heavy batteries and rare earth magnets, and in the ongoing search for a cleaner way to fix the nitrogen that feeds the planet.

Chemistry, for all its reputation as a subject of dramatic color changes and explosive demonstrations, turns out to do its most consequential work quietly, in concentrations too small to see, in reactions too subtle to photograph for a science museum display. The story of trace elements, told across the fourteen chapters of this book, is ultimately a story about paying attention to that quiet work, about recognizing that the elements too small to notice are very often the elements too important to ignore, and about the scientists, chemists, physicians, metallurgists, and geochemists, some celebrated with Nobel Prizes and some, like Harry Brearley, met largely with institutional indifference in their own lifetimes, who spent their careers learning to listen closely enough to hear what a whisper of a few parts per million was actually trying to tell them.

That work continues. Somewhere right now, a researcher is testing a new catalyst formulation that might reduce the energy cost of feeding the world. Somewhere, a materials scientist is running a computational simulation searching for a magnet that does not require dysprosium. Somewhere, a battery chemist is trying to build a cathode that works just as well without cobalt. The specific elements and the specific problems will keep changing, as they always have. The underlying pattern, small quantities exerting outsized influence, waiting to be discovered, understood, and carefully, deliberately deployed, seems likely to remain exactly what it has been for the past seven thousand years of human material history: the quiet, persistent, and remarkably durable alchemy of almost nothing.

The Weight of a Whisper

There is a particular kind of vertigo that can set in once you have spent enough time thinking about parts per million, and it tends to arrive somewhere in the ordinary business of an ordinary day. You might notice it while salting a pot of water for pasta, catching yourself thinking, unbidden, about a Michigan grocery shelf in the spring of 1924. You might notice it while your phone silently charges overnight on a nightstand, the lithium inside its battery shuttling back and forth between two electrodes a chemist in Texas spent years perfecting. You might notice it while running a stainless steel knife under the kitchen tap, or while sitting in traffic behind a car whose exhaust, however imperfectly, has already been stripped of the worst of its poison by a few grams of metal you will never see.

This is not a bad kind of vertigo to develop. It is, in fact, close to the entire purpose of this book. The world is dense with this sort of quiet leverage, elements and compounds working at concentrations too small for any of our native senses to register, and once you start noticing the pattern in one place, you tend to start noticing it everywhere, which changes, in small but genuine ways, how the built and grown world around you actually reads.

It changes how you think about scarcity, for one thing. The chapters of this book have moved between elements that are, in an absolute geological sense, genuinely rare, rhodium and dysprosium among them, and elements that are, by any reasonable measure, almost embarrassingly abundant, iodine and iron and the carbon that makes up every living thing on Earth. What determines whether an abundant element actually reaches the people or processes that need it turns out to have less to do with how much of it exists in the earth's crust and more to do with something considerably more human: whether anyone has bothered to look for the deficiency, whether anyone has built the supply chain, whether anyone has done the patient, often unglamorous work of figuring out exactly how much is needed, exactly where, and exactly in what chemical form. Molybdenum is not scarce. For decades, the knowledge that certain tropical soils needed it was scarce. Zinc is not scarce. For most of human history, the recognition that a diet built almost entirely around unleavened wheat bread could starve a growing body of it was scarce. The lesson, if this book has one single lesson to offer beyond the specific stories it tells, may be exactly that: attention is very often the truly limiting reagent, not the element itself.

It also changes how you think about risk. This book has tried, deliberately and I hope successfully, not to tell a uniformly triumphant story. The chapter on tetraethyl lead sits where it does, roughly at this book's midpoint, because it needed to sit somewhere that would not let the reader settle into an easy, uncomplicated admiration for trace element chemistry as an unambiguous force for good. The same fundamental principle, a small quantity of a well-chosen element producing an outsized effect on the systems around it, does not come pre-loaded with any guarantee about which direction that effect will run. Thomas Midgley Jr. deployed exactly the kind of chemical cleverness this book has celebrated in a dozen other contexts, and the result was decades of preventable harm to the developing brains of children who never had any say in the matter. The chemistry itself offered no warning. The warning had to come from people, from Clair Patterson's stubborn insistence on following his data even when it made him deeply unpopular with a powerful industry, and from the slow, difficult work of translating scientific findings into actual regulatory change against sustained and well-funded resistance.

There is a related pattern worth naming explicitly, one that runs quietly beneath several of this book's chapters without ever becoming their central subject: the gap between when a trace element's danger or benefit first becomes scientifically clear and when that knowledge actually reaches the policies, products, and practices that could act on it. Marine and Kimball's Akron trial demonstrated iodine's protective effect conclusively by 1920, yet iodine deficiency persisted in significant pockets of the developed world for decades afterward and remains a documented public health concern for a meaningful share of the global population today. Ananda Prasad's zinc deficiency research faced a full sixteen years of institutional skepticism before receiving official recognition. Clair Patterson's lead contamination findings took three decades to translate into a complete regulatory phase-out. This lag is not a footnote to the history of trace element science. It is very nearly the main plot, and it suggests that the harder and more consequential work, in many of these stories, was never really the initial chemistry. It was the sustained, unglamorous labor of getting a skeptical world to act on what the chemistry had already made clear.

And it changes, perhaps most usefully, how you think about the future. The final chapter of this book left several genuinely open questions hanging rather than resolved: whether rare earth magnet dependence can be meaningfully reduced through cerium-cobalt alternatives or aggressive recycling, whether electrochemical nitrogen fixation will ever mature into a genuine replacement for the century-old Haber-Bosch process, whether the world's growing hunger for cobalt can be satisfied without perpetuating the labor conditions that have drawn sustained international scrutiny to parts of the mining sector in the Democratic Republic of the Congo. These are not questions this book, or any book, can answer in advance. They are live research problems, being worked on right now, in laboratories and mines and legislative chambers around the world, by people continuing a tradition that stretches back, as the thirteenth chapter of this book tried to show, to an anonymous metalworker in the ancient Near East who first noticed that a particular kind of copper ore produced an unusually hard and useful metal.

What I hope this book leaves you with, beyond the specific facts and stories it has tried to tell accurately and tell well, is a slightly recalibrated sense of where to look when trying to understand how the material world actually works. Not always at the largest, most visible ingredient in a system, the bulk steel, the bulk fuel, the bulk grain, but sometimes, and very often at exactly the moments that matter most, at the whisper hiding just beneath it: the pinch, the trace, the part per million that was, all along, quietly doing most of the work.

The stories told across this book draw on published scientific literature, historical archives, biographical accounts, and contemporary reporting spanning more than two centuries of chemistry, medicine, metallurgy, and materials science. Where specific dates, measurements, and quotations appear in the text, they reflect the documented historical and scientific record as established by the researchers, journalists, and institutions who did the original work of discovering, measuring, and recording them.

Readers interested in exploring any of this book's fourteen central stories in greater depth will find a rich body of further reading available. The early history of the Haber-Bosch process and its role in twentieth century agriculture has been examined in considerable depth by the geographer and policy analyst Vaclav Smil, whose scholarship on the topic remains a standard reference. Dorothy Hodgkin's own scientific papers and subsequent biographical treatments offer a fuller account of the eight-year effort to determine vitamin B12's structure than this book's chapter could accommodate. The history of tetraethyl lead and Clair Patterson's research has been the subject of extensive investigative journalism and environmental history scholarship, much of it drawing on Environmental

Protection Agency records, congressional testimony, and Patterson's own published research from the 1960s onward. Contemporary developments in battery chemistry, rare earth supply chains, and green ammonia research continue to be actively documented in the ongoing scientific literature, and readers interested in the current state of any of these still-evolving fields are encouraged to consult recent publications from the relevant scientific journals and international research institutions, since the specific details of active, ongoing research inevitably continue to develop past the point where any printed book can remain fully current.

Every effort has been made to ensure the accuracy of the historical dates, scientific findings, and biographical details presented throughout this book. As with any work synthesizing a broad span of scientific and historical material, readers pursuing rigorous academic or professional research on any of these topics are encouraged to consult primary sources and current peer-reviewed literature directly.

A note on scope is also worth including here. Fourteen chapters, however comprehensive each has tried to be, cannot cover every trace element story that deserves telling. Selenium's role in thyroid and immune function, manganese's function in bone development and its more recently studied neurological effects at high occupational exposure, fluoride's contested and much debated role in dental public health, and iridium's use in identifying the asteroid impact associated with the extinction of the dinosaurs all represent genuinely rich additional material that a longer book, or a future one, might well take up. The fourteen stories chosen for this volume were selected not because they exhaust the subject but because, taken together, they illustrate the book's central argument with unusual clarity and across an unusually wide range of human domains, spanning agriculture, medicine, industry, environmental history, and the deep chronology of archaeology and geology. Readers who find themselves, by the book's final page, curious about the trace elements it did not have room to cover should take that curiosity as a sign that the book has done its job.

Priya Anantharaman writes about the hidden chemistry shaping everyday life, with a particular fascination for the small causes behind large consequences. Her work explores the intersection of scientific history and material culture, tracing how obscure laboratory discoveries quietly reshape agriculture, medicine, industry, and technology. She approaches each subject with a researcher's insistence on verified fact and a storyteller's instinct for narrative. This is her latest exploration of chemistry's quietest and most consequential actors.

www.ingramcontent.com/pod-product-compliance
Lightning Source LLC
LaVergne TN
LVHW100143061126
851857LV00010B/217